PERGAM… …ARY

of Science, Techno… …Studies

The 1000-volume original paperback library in aid of edu…

industrial trainir…

Pub…

Principles of Industrial Microbiology

Principles of Industrial Microbiology

by

ALAN RHODES
Ph.D. (Cambridge), D.Sc. (Leeds), F.I.Biol.

and

DEREK L. FLETCHER
B.Sc. (London), C.Eng., A.M.I.Chem.E.

PERGAMON PRESS

OXFORD · NEW YORK · TORONTO
SYDNEY · PARIS · FRANKFURT

U.K. Pergamon Press Ltd., Headington Hill Hall, Oxford OX3 0BW, England

U.S.A. Pergamon Press Inc., Maxwell House, Fairview Park, Elmsford, New York 10523, U.S.A.

CANADA Pergamon of Canada Ltd., 75 The East Mall, Toronto, Ontario, Canada

AUSTRALIA Pergamon Press (Aust.) Pty. Ltd., 19a Boundary Street, Rushcutters Bay, N.S.W. 2011, Australia

FRANCE Pergamon Press SARL, 24 rue des Ecoles, 75240 Paris, Cedex 05, France

WEST GERMANY Pergamon Press GmbH, 6242 Kronberg-Taunus, Pfeerdstrasse 1, Frankfurt-am-Main, West Germany

---

First edition 1966

Reprinted 1975, 1977

Library of Congress Catalog Card No. 66-21144

Printed in Great Britain by A. Wheaton & Co., Exeter

ISBN 0 08 011905 0

Contents

List of Tables

List of Figures

Plates

List of Abbreviations

RNA	ribonucleic acid
DNA	deoxyribonucleic acid
mRNA	messenger ribonucleic acid
sRNA	soluble or transfer ribonucleic acid
ribosomes	ribonucleoprotein particles
Ⓟ	orthophosphate radical
P_i	orthophosphate
PP_i	pyrophosphate
$\sim$	energy-rich bond
acetyl$\sim$S.CōA	acetyl-coenzyme A or active acetate
ATP	adenosine triphosphate
ADP	adenosine diphosphate
UTP	uridine triphosphate
UDP	uridine diphosphate
UDP-glucose	uridine diphosphate glucose
GTP	guanosine triphosphate
GDP	guanosine diphosphate
ITP	inosine triphosphate
IDP	inosine diphosphate
TPP	thiamine pyrophosphate, aneurin diphosphate, co-carboxylase, or the pyrophosphate of vitamin B_1
α-hydroxyethyl-TPP	α-hydroxyethyl-thiamine pyrophosphate or active acetaldehyde
NAD or $NAD^{\oplus}$	oxidized form of nicotinamide adenine dinucleotide
$NADH_2$ or $NADH + H^{\oplus}$	reduced form of nicotinamide adenine dinucleotide
NADP or $NADP^{\oplus}$	oxidized form of nicotinamide adenine dinucleotide phosphate
$NADPH_2$ or $NADPH + H^{\oplus}$	reduced form of nicotinamide adenine dinucleotide phosphate
FAD	oxidized form of flavin adenine dinucleotide
$FADH_2$	reduced form of flavin adenine dinucleotide
EDTA	ethylene-diamine-tetracetic acid
B.O.D.	biological oxygen demand

psia	pounds per square inch absolute (i.e. the absolute pressure)
psig	pounds per square inch gauge (i.e. the pressure, as measured by the gauge, above atmospheric pressure)

CHAPTER 1

Introduction

1. Historical background

Many of man's domestic processes of great antiquity depended on fermentative change: they include conversion of grapes to wine, wine to vinegar and honey to mead; the retting of flax; the procedures for brewing and cheese-making. These practices were recognized to be part of life: they were accepted with little thought about the mechanisms involved, just as the fermentative disposal in soil of household waste was taken for granted.

Micro-organisms were first described as "very little animalcules" by Antonie van Leeuwenhoek in his letters to the Royal Society, written in 1674–5. It was not until 1861, however, that Louis Pasteur was able to refute the "theory of spontaneous generation" by demonstrating the dependence of fermentations on the activity of microbes that entered infusions from the air.

Pasteur's subsequent work showed that fermentations characterized by formation of particular products (e.g. acetic acid, alcohol, butyric acid, lactic acid) were dependent on different kinds of micro-organisms and that contamination by undesirable microbes resulted in "bad" fermentations.

It was assumed for some time that microbes were characteristically degradative in action, but in the years between 1930 and 1960 Raistrick and his colleagues at the London School of Hygiene and Tropical Medicine demonstrated the phenomenal capabilities of micro-organisms for biosynthesis from simple substrates of complex organic compounds often in high yield.

Previously, towards the end of the nineteenth century, the foundations of biochemistry had been laid when the Buchner brothers demonstrated that yeast juice freed from living cells could bring about the alcoholic fermentation of cane-sugar solution.

Another step of great potential importance was the discovery by Fleming at St Mary's Hospital, London, in 1928 that a common fungus, saprophytic on fruit, produced in culture a substance that inhibited the growth of certain bacteria pathogenic to man. The antibiotic was difficult to extract from the mould cultures, so that another 12 years elapsed before Florey, Heatley and Chain extracted the clinically effective ingredient now known as penicillin.

The need for penicillin in the Second World War fortunately coincided with great advances in techniques of chemical engineering. The problems inherent in large-scale, submerged, aerobic culture of micro-organisms were quickly solved, with the result that commercial production in large, mechanically agitated and aerated fermenters began in U.S.A. in 1942. This led quickly to a similar development in the United Kingdom.

Waksman's discovery in 1944 of streptomycin, another clinically valuable antibiotic, demonstrated that useful new antibiotics could be found by deliberate investigation of soil micro-organisms for ability to produce these metabolites. Similar techniques revealed such a large number of new antibiotics that within the next 10–15 years the treatment of human disease was revolutionized.

The development of industrial microbiology may be seen to have passed through three distinct phases. At first, fermentations giving a variety of characteristic foods and beverages were dependent upon indigenous micro-organisms. Recognition of the essential role in fermentations of micro-organisms, availability of techniques for selected pure cultures and knowledge of the benefits to be gained by sterile or aseptic procedures together permitted in the second phase production of relatively simple products (e.g. commercial yeast, glycerol, industrial alcohol, butanol–acetone) by large-scale industrial fermentation processes having little or no requirement for aeration. The third phase in the progress of industrial microbiology, initiated by development of penicillin, is characterized by the intensive and purposive search for soil micro-organisms able to produce valuable new antibiotics, and by the benefits gained when advances in chemical engineering were applied to microbiological problems.

Up to this time, the industrial processes depended largely upon empirical microbiological techniques, but industrial microbiology is now on the threshold of a new era that may permit rationalization of many processes in terms of the underlying biological, chemical and engineering principles.

2. Purpose of the book

Information relating to the most recent improvements in industrial microbiological processes is not accessible, either because the relevant information is withheld as "know-how", or because the discoveries are the subject of current patent applications. It is therefore inevitable that a considerable delay must occur between the time of introducing to industry new procedures and the time when this information is disclosed to the public; that is why no book recounting details of industrial microbiological processes can ever hope to be up to date.

We have, therefore, tried to look at industrial microbiology from the point of view of its underlying principles in chemical engineering, biochemistry and microbiology, and this has led to a different arrangement of subject matter from that normally adopted when processes are grouped according to the chemical nature of their products.

This is a book intended for biologists, chemists and engineers interested in the problems of industrial microbiology and for students intending to enter the industry. Many sciences are utilized in industrial microbiological processes, and it is seldom found that a single discipline is adequate for solving a particular problem: success usually depends upon fusion of ideas, and this is not possible unless all are familiar with common basic principles.

Suggested books of general reference

1. BADGER, W. L. and McCABE, W. L., *Elements of Chemical Engineering*, 2nd edn., McGraw-Hill, New York and London, 1936.
2. BALDWIN, E., *Dynamic Aspects of Biochemistry*, 3rd edn., Cambridge University Press, 1957.
3. BERNFELD, P. (ed.), *Biogenesis of Natural Compounds*, Pergamon, Oxford, London, New York and Paris, 1963.
4. CRUICKSHANK, R. (ed.), *Mackie and McCartney's Handbook of Bacteriology*, 10th edn., Livingstone, Edinburgh and London, 1960.
5. COULSON, J. M. and RICHARDSON, J. F., *Chemical Engineering*, 2nd edn., (2 vols.), Pergamon Press, London, 1956.
6. FLORKIN, M. and STOTZ, E. H. (eds.), *Group-transfer Reactions*, Vol. 15, *Comprehensive Biochemistry*, Elsevier, Amsterdam, London and New York.
7. FRAZIER, W. C., *Food Microbiology*, McGraw-Hill, New York, Toronto and London, 1958.
8. GREENBERG, D. M. (ed.), *Metabolic Pathways*, Vols. I and II, Academic Press, New York and London, 1960 and 1961.
9. GUNSALUS, I. C. and STANIER, R. Y. (eds.), *The Bacteria: a treatise on structure and function*, Vol. I *Structure*, 1960; Vol. II *Metabolism*, 1961; Vol. III *Biosynthesis*, 1962; Vol. IV *Physiology of Growth*, 1962. Academic Press, New York and London.
10. HARRIS, M., *Pharmaceutical Microbiology*, Baillière, Tindall & Cox, London, 1964.
11. HAWKER, L. E., LINTON, A. H., FOLKES, B. F. and CARLILE, M. J., *Biology of Micro-organisms*, Edward Arnold, London, 1960.
12. INGRAHAM, L. L., *Biochemical Mechanisms*, John Wiley, New York and London, 1962.
13. KARLSON, P., *Introduction to Modern Biochemistry*, Academic Press, New York and London, 1963.
14. LONG, C. (ed.), *Biochemists' Handbook*, E. & F. Spon, London, 1961.
15. MILLER, M. W., *The Pfizer Handbook of Microbial Metabolites*, McGraw-Hill, New York, Toronto and London, 1961.
16. PENSO, G. and BALDUCCI, D., *Tissue Cultures in Biological Research*, Elsevier, Amsterdam, London and New York, 1963.
17. PERRY, J. H. (ed.), *Chemical Engineers' Handbook*, 3rd edn., McGraw-Hill, New York and London, 1950.
18. PRESCOTT, S. C. and DUNN, C. G., *Industrial Microbiology*, 3rd edn., McGraw-Hill, New York, Toronto and London, 1959.
19. RAINBOW, C. and ROSE, A. H. (eds.), *Biochemistry of Industrial Micro-organisms*, Academic Press, London and New York, 1963.
20. ROSE, A. H., *Industrial Microbiology*, Butterworths, London, 1961.
21. SMITH, G., *An Introduction to Industrial Mycology*, 5th edn., Edward Arnold, London, 1960.
22. SPANNER, D. C., *Introduction to Thermodynamics*, Academic Press, London and New York, 1964.
23. STEEL, R. (ed.), *Biochemical Engineering*, Heywood, London, 1958.

24. STODOLA, F. H., *Chemical Transformations by Micro-organisms, E. R. Squibb Lectures on Chemistry of Microbial Products*, John Wiley, New York, 1958.
25. UNDERKOFLER, L. A. and HICKEY, R. J., *Industrial Fermentations*, Vols. I and II, Chemical Publishing Co., New York, 1954.
26. VILBRANDT, F. C., *Chemical Engineering Plant Design*, 3rd edn., McGraw-Hill, New York and London, 1959.
27. WAKSMAN, S. A., *Soil Microbiology*, John Wiley, New York, 1952.
28. WAKSMAN, S. A., *The Actinomycetes*. I. *Nature, Occurrence and Activities*, 1959. II. *Descriptions of Genera and Species of Actinomycetes*, 1961. III. *Antibiotics of Actinomycetes*, 1962. Baillière, Tindall & Cox, London.
29. WALKER, W. H., LEWIS, W. K., MCADAMS, W. H. and GILLILAND, E. R., *Principles of Chemical Engineering*, 3rd edn., McGraw-Hill, New York and London, 1937.
30. WEBB, F. C., *Biochemical Engineering*, D. van Nostrand, London, Toronto and New York, 1964.
31. WHITMARSH, J. M., *British Fermentation Industries*, Sir Isaac Pitman, London, 1958.

Suggested annual reviews

1. CLIFTON, C. E. (ed.), *Annual Review of Microbiology*, Annual Reviews, U.S.A.
2. DREW, T. B., HOOPES, J. W., Jr. and VERMEULEN, T. (eds.), *Advances in Chemical Engineering*, Academic Press, New York and London.
3. HOCKENHULL, D. J. D. (ed.), *Progress in Industrial Microbiology*, Heywood, London.
4. LUCK, J. M. (ed.), *Annual Review of Biochemistry*, Annual Reviews, U.S.A.
5. MACHLIS, L. (ed.), *Annual Review of Plant Physiology*, Annual Reviews, U.S.A.
6. NORD, F. F. (ed.), *Advances in Enzymology*, Interscience, New York.
7. UMBREIT, W. W. (ed.), *Advances in Applied Microbiology*, Academic Press, New York and London.

CHAPTER 2

Design of a Fermentation Unit

1. Essential features of a fermentation unit

The fermentation unit in industrial microbiology is analogous to the chemical plant in chemical industry. When differences arise, they are due to the requirements of the biological process for sterility and to the use of its constituent enzymic reactions instead of chemical reactions aided by inanimate catalysts sometimes operating at elevated temperature and pressure.

The principles of design are generally applicable regardless of scale of operation: the description below relates to a fermentation unit suitable for a small-scale experimental unit, to a pilot plant intended to provide experimental quantities of mould metabolites and to a full-scale production plant. The general lay-out is illustrated in Fig. 2.1.

The *fermenters* must be designed correctly and operated successfully before the ancillary units can function efficiently. The capital invested in fermenters and their immediate services is large, so that only minor modifications are feasible once they are installed.

The *culture collection laboratory* (see Chapter 4) provides suitable inocula for initiation of the required microbiological processes and makes routine sterility checks on the fermentation broths.

The *control laboratory* monitors each fermentation process with assays for reducing sugars, total hydrolysed sugars and available ammoniacal nitrogen, together with routine determinations of pH and mycelial weight, in order to provide a biochemical picture of the progress of each batch. Fermentation and extraction yields are

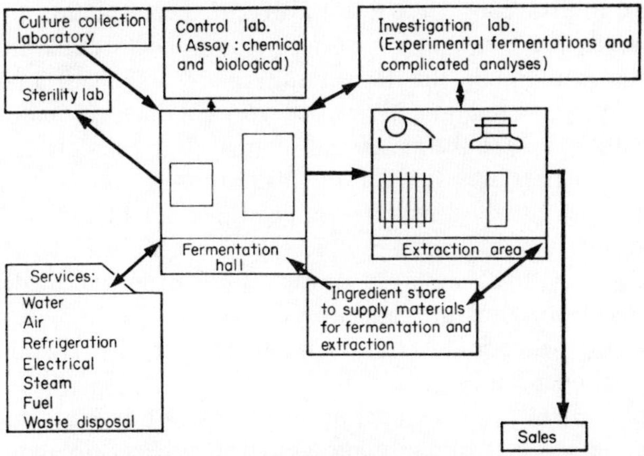

FIG. 2.1. General lay-out for a fermentation unit.

determined by chemical or biological analyses of samples taken at various points in the process to construct mass balances: this information permits identification of the stages where losses occur.

The *services* needed to operate a number of fermenters are numerous.

(1) Clean water for "batching" the fermentation media.
(2) Cooling water for temperature control, treated to remove hardness and to inhibit microbial growth.
(3) Sterile compressed air for aeration.
(4) Boiler house to supply sufficient high pressure steam for sterilization of ingredients.
(5) Adequate supply of electricity to run stirrer motors and air-compressors.
(6) Acceptable system of waste disposal.

The *ingredient store* needs to provide bulk storage for diverse materials, including inflammable solvents, sugars, cornsteep liquor, corn meals and ion exchange resins.

The purpose of the *investigation laboratory* is to conduct microbiological and biochemical studies of the fermentation process and to consider the chemistry of the extraction process employed, with a view to increasing yields in the "broth" (nutrient medium containing the fermenting micro-organism) and the efficiency of extraction. Facilities for conducting small-scale fermentations in shake-flasks and in small (for example 5, 10 or 15 l.) fermenters are essential for this purpose.

The *extraction area* must be equipped to extract from the broth all the variety of industrial products that need to be isolated. The extraction method employed on any given product will depend upon its chemical nature, but the method will need to be adapted to one of the major extraction methods generally available to industrial microbiology.

The *sales unit* has to sell the product itself or novel information about the conduct of the microbiological process or about the extraction and purification of the desired product.

2. Systems of measurement

Industrial microbiological processes have world-wide distribution and all require the joint efforts of biologist, chemist and engineer. It is therefore important that the systems of measurement employed should be internationally interchangeable. Confusion rarely arises between the metric and British systems of measurement, because the difference between the litre and the gallon, on the one hand, and between the metre and the foot, on the other, is evident. Difficulties may arise, however, if the scale of temperature measurement is not specified as either Centigrade or Fahrenheit. Complete ambiguity may develop between the American and British systems because the U.S.A. gallon is approximately four-fifths of the British gallon, the American pint is four-fifths of the British pint and the American ton (short ton) contains 2000 lb instead of the 2240 lb of the British ton (long ton).

Errors may also occur when abbreviations for units of measurement are still further abbreviated: pounds per square inch gauge (psig) is the pressure, as measured by the gauge, above atmospheric pressure at that particular time, but when shortened to psi there is nothing to indicate whether the absolute pressure (psia) or the gauge pressure is intended. We therefore note that

$$\text{psia} = \text{psig} + \text{atmospheric pressure}$$

Table 2.1. shows equivalent values for length, volume, mass, energy, pressure and temperature in the systems of measurement encountered in industrial microbiology.

TABLE 2.1. TABLE OF EQUIVALENTS IN DIFFERENT SYSTEMS OF MEASUREMENT

Length	Volume	Mass
1 m	1 l.	1 kg
100 cm	1000 ml	1000 g
1000 mm	0·2642 U.S.A. gal	2·205 lb
39·37 in.	0·2201 Imp. gal	0·0009842 long tons
3·28 ft	0·03531 ft^3	0·001102 short tons
10^6 μ		

Energy	Pressure	Temperature
1 Calorie (kilogram-calorie = 1000 calories)	1 atm	100°C
	14·696 lb/in^2	212°F
3·9657 B.t.u.	1·0332 kg/cm^2	373·18°K
0·001162 kWh	29·92 in. Hg at 0°C	Temperature of boiling water
4184·0 J	33·93 ft H_2O at 15°C	
0·001558 h.p.-hr.		

It is also convenient to memorize a few other relationships. Degrees Centigrade are converted to degrees Fahrenheit by the equation:

$$°F = \frac{9°}{5} C + 32,$$

and to degrees Kelvin by the equation:

$$°K = °C + 273{\cdot}18.$$

Finally, the metric system specifies that 1 l. of water at 4°C weighs 1 kg; in the British system 1 imperial gallon of water at 62°F weighs 10 lb; in the American system 1 pint (8 U.S.A. pints to the U.S.A. gallon) weighs 24/25 (0·96) lb.

3. Fermenter design

The function of a fermenter is to provide an environment optimal for each particular microbiological process. There are two types of fermenters: aerobic and anaerobic. Differences in design between the two are mainly determined by the amounts of aeration and agitation required.

3.1. *Fermenters for aerobic fermentations*

This type of fermenter (see Fig. 2.2.) is a closed vessel that can be sterilized, aerated, stirred and have the temperature of its contents regulated with a high degree of accuracy. The shape is usually cylindrical with a rounded base and a smooth interior to facilitate cleaning. The height is 50–100% greater than the diameter. Regularly spaced baffles, usually four in number, are set

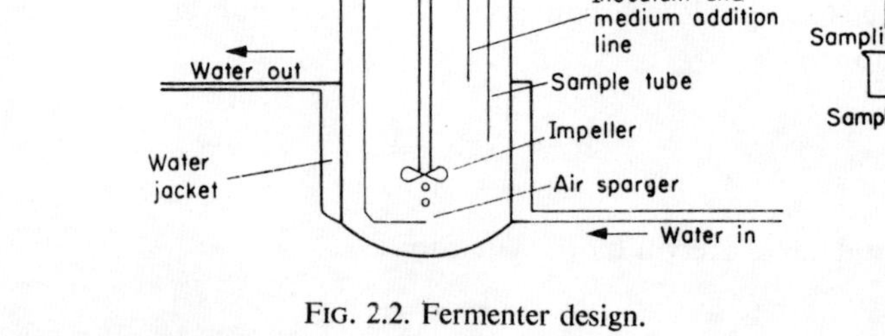

FIG. 2.2. Fermenter design.

down the length of the fermenter perpendicular to its walls. The stirrer shaft runs through the centre of the fermenter and carries one or more impellers, according to the height of the fermenter and the intensity of agitation required. The stirrer shaft and motor are designed to permit absorption of up to 0·9 h.p./100 gal fermentation medium, and the cooling system is designed to remove both the heat generated by stirring and the metabolic heat generated by the microbial culture, while maintaining at optimum temperature the fermentation liquor: 25°C for the penicillin fermentation, 28°C for streptomycin, and 32°C for production of 2,3-butyleneglycol.

Air sterilized by filtration passes through a non-return valve at a maximum rate equivalent to one volume of medium per minute to enter the fermentation liquor by an air sparger at the bottom of the fermenter, located just beneath the lowest impeller. Design of the air sparger is a compromise between having the sparger holes sufficiently large to present minimum resistance to the air flow and sufficiently small to prevent clogging of the apertures by particles of medium. The air is finally exhausted to the atmosphere through a long pipe that is periodically sterilized. Contamination of other fermenters by the spent air is reduced to a minimum by exhausting to leeward of the fermenter air intake.

In some industrial processes a bacteriological filter is included in the exhaust pipe to minimize dissemination of the micro-organisms present in the spent air.

Fermentation broths may foam at intervals: the operator controls the level of foam by addition of anti-foam oils so that broth is not lost as foam. Any inadvertent blow-over of broth in the exhaust air may be separated by a cyclone and harvested.

A sample tube, closed by two external valves, *A* and *B* in series, extends into the vigorously agitated liquor to permit sampling of the medium. A steam-line under pressure is connected between these two valves, thus applying to this section of pipe a steam seal that enables it to be sterilized after every sample is withdrawn, and provides a heat barrier against the entry of contaminating organisms.

This same principle of the steam seal is used to permit other sterile connections to the fermenter for the purpose of addition of nutrient, introduction of inoculum, addition of anti-foam or transfer of broth from one vessel to another.

Inoculum prepared by the methods described in Chapter 4 is transferred to a small development fermenter, which has approximately 10% of the volume but is of the same general basic design as the large fermenter already described in Fig. 2.2. When the desired amount of microbial cells or mycelium has been grown, the contents of the development fermenter are transferred to the main fermentation vessel by air pressure through a sterilized transfer-line, the operation being completed as rapidly as possible to minimize any harmful effect of suboptimal aeration.

The sterile transfer is effected in the manner indicated in Fig. 2.3. The valves *C* and *D* are set to connect the two vessels after first closing valves *E* and *F*. Steam enters the transfer-line and passes to the drains through the partially opened valves *G* and *H*; the pressure is allowed to remain at 15 psig for 30 min, thus

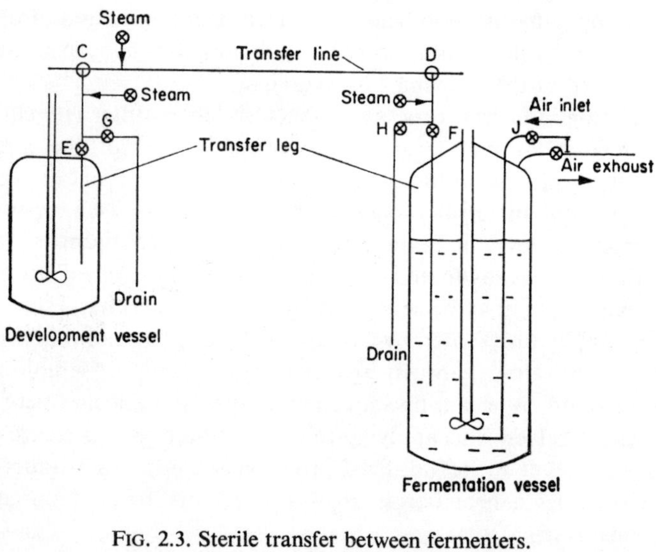

FIG. 2.3. Sterile transfer between fermenters.

sterilizing the transfer-line. Pressure is raised in the fermentation vessel by closing the air-exhaust valve *I*, and sterilized cold medium is forced back to drain through the line on the development vessel when the steam supply is cut off and valve *F* is opened: this procedure cools the transfer-line and legs. The drain line of the development vessel is closed by operation of valve *G*. Pressure is then lowered in the fermentation vessel by controlled simultaneous opening of the air inlet *J* and the exhaust valve *I*, pressure in the development vessel being raised in the usual manner. Valve *E* is then opened to permit the whole contents of the development vessel to be transferred to the fermentation vessel. Valve *F* is closed, and the transfer-line is finally sterilized again by steam, so that it is ready for another transfer.

3.2. *Fermenters for "anaerobic" fermentations*

Basically a fermenter designed to operate under micro-aerophilic ("anaerobic") conditions will be the same as that designed to operate under aerobic conditions, except that design features for intense agitation and aeration are unnecessary. The intensity of aeration need only be sufficient for good growth of cells and the intensity of agitation sufficient for mixing and maintenance of temperature. Many anaerobic fermentations do, however, require an initial aerobic growth phase (e.g. dextran, alcohol, 2,3-butylene-glycol): it is only the later or productive part of the fermentation that needs reducing or micro-aerophilic conditions. Completely anaerobic microbiological processes are rare in industry; the fermentation of the tetanus bacterium is a familiar example.

4. Extraction equipment

4.1. *Filters*

Mould metabolites are either soluble or insoluble in the broth, and they may be present either in the mycelium or released into the surrounding fluid. The first stage of extraction invariably involves some sort of clarification or filtration to remove insoluble material or the fermenting organism or both. The most versatile but costly piece of equipment for this purpose is the rotary vacuum filter, illustrated in Fig. 2.4. This may be operated either as a pre-coat filter for broths difficult to filter (e.g. chlortetracycline) or as a string-discharge filter for broths relatively easy to filter (e.g. griseofulvin).

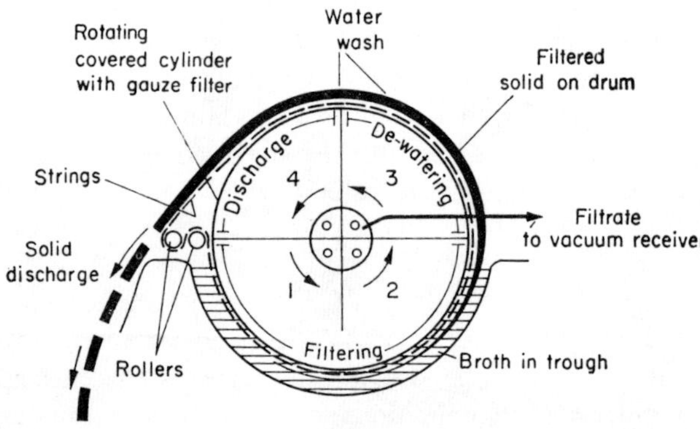

FIG. 2.4. The rotary vacuum filter operating by string discharge.

The filter consists of a rotating cylinder covered with a gauze filter (metal or fabric), and this is partially immersed in a trough containing the broth to be filtered. The interior of the drum can be regarded as divided into four compartments to which vacuum may be applied, either to all the compartments simultaneously or to only the first three. To operate as a pre-coat filter, a slurry (1–5% w/v) of filter-aid (e.g. kieselguhr or dicalite) in water is pumped into the trough, and vacuum is applied to compartments 1, 2, 3 and 4 while the drum slowly rotates (about 1 rev/min) anticlockwise. The filter-aid is retained on the gauze filter as a bed of uniform depth, but the water is sucked through into a vacuum receiver. When a sufficient thickness (1–2 in.) of filter-aid has been deposited on the surface of the cylinder, the slurry is replaced by fermentation broth. Solid particles picked up on the bed at positions 1 and 2 remain on the surface of the filter-aid, but filtrate is separated by entering the drum and proceeding to the vacuum receiver. A water wash may be applied as the solids pass position 3. At position 4 a sharp knife inserted tangentially to the motion, but just clear of the surface of the filter-aid, cuts the thin layer of broth solids from the surface of the filter.

To operate as a string-discharge filter, long lengths of string are threaded around the drum passing over two rollers, as shown by the dotted lines in Fig. 2.4. The solids accumulated on the surface of the drum by suction are lifted free by the strings as they pass over the rollers. This operation is facilitated if the vacuum is released in compartment 4.

The plate-frame filter is relatively cheap and is suitable for fermentation broths with low solids content and a correspondingly low resistance to filtration (e.g. yeast fermentation broths). Filter cloth is stretched over a rectangular plate, vertically ribbed to ensure passages between the surface of the plate and of the cloth. Pairs of plates, each separated by a frame, are tightly assembled in series, so that a watertight compartment, bounded by the frame, is formed between each two plates. Broth under pressure is led into each watertight compartment of the plate-frame filter, but the outlets, one to each plate, are so placed that exit is possible only by passing through the filter cloth. Filtrate may be collected if required, but solids are retained to accumulate in each watertight compartment of the filter.

4.2. *Centrifuges*

Products remaining in the filtrate may be amenable to solvent extraction and separation by centrifuge. The feasibility of this procedure will depend upon the partition coefficient.

Suppose a substance P dissolved in water is mixed with an equal volume of an immiscible solvent, distribution of P between the solvent phase and the aqueous phase will be determined by the equation:

$$k = \frac{[P_s]}{[P_a]} = \text{partition coefficient},$$

where $[P_s]$ is the concentration at equilibrium of P in the solvent phase and $[P_a]$ the corresponding concentration in the aqueous phase. The value of k will depend upon a number of factors, including the chemical nature of the product, the solvent used, the temperature and the pH. For organic acids or bases, the partition coefficient also depends on the degree of ionization, so that the value of k will be largely governed by pH. The ionic form is usually soluble in the aqueous phase, but the non-ionized molecule is soluble in the organic phase, so that the corresponding partition coefficients favour the aqueous and organic phases respectively:

$$\text{(i)} \qquad \underset{\text{(soluble in organic phase)}}{\text{H Acid}} \underset{\text{acid}}{\overset{\text{alkali}}{\rightleftharpoons}} \underset{\text{(soluble in aqueous phase)}}{H^{\oplus} + \text{acid}^{\ominus}}$$

$$\text{(ii)} \qquad \underset{\text{(soluble in organic phase)}}{\text{Base OH}} \underset{\text{alkali}}{\overset{\text{acid}}{\rightleftharpoons}} \underset{\text{(soluble in aqueous phase)}}{\text{base}^{\oplus} + OH^{\ominus}}$$

The extraction of penicillin, an acidic substance, illustrates this phenomenon. The broth filtrate is mixed with a smaller volume of butyl acetate and then acidified with phosphoric acid, so that the non-ionized penicillin concentrates in the solvent layer, thus permitting separation from the aqueous layer. The penicillin-rich butyl acetate phase is then mixed with phosphate buffer at pH 7·0, so that the ionized sodium salt of penicillin is now concentrated in an even smaller volume of aqueous phase, from which the antibiotic is finally isolated.

Solvents may be separated from water by standing, but it is usual to hasten the process by increasing with centrifugal motion the separating force.

The centrifugal force is given by the expression $W\omega^2 r/g$, where W is the weight of the material being rotated with an angular velocity ω at a radial distance r from the centre of motion, and g is the acceleration due to gravity (980·7 cm/sec²).

A typical centrifugal separating machine leads the solvent mixture into the centre of a rotating bowl, as shown in Fig. 2.5. The mixture is thrown by centrifugal force to the peripheral walls of the bowl, where it separates into a heavy phase and a light phase.

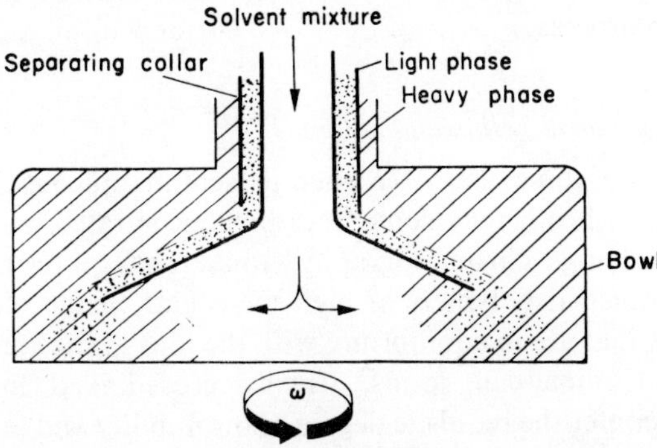

FIG. 2.5. Vertical section through a typical bowl centrifuge.

Continued admission of solvent mixture to the bowl forces the two phases into the neck of the machine, where suitable placement of the collar permits their separation and collection from different outlets.

The partition coefficient and volume ratio will determine the amount of extractable material lost in the spent fluid from a single extraction. Some of this loss may be recovered by insertion of a second stage, whereby the spent liquor is extracted with fresh solvent. The half-rich solvent from this second stage may then be coupled for extraction of the neat broth of the preceding stage, as shown in Fig. 2.6.

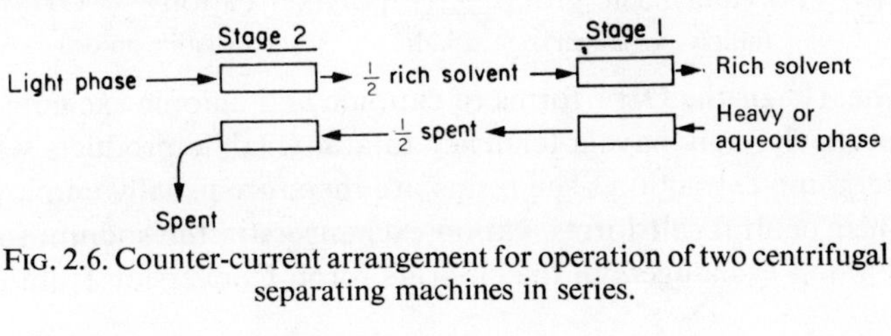

FIG. 2.6. Counter-current arrangement for operation of two centrifugal separating machines in series.

The increased efficiency that results from inclusion of the second extraction stage may be appreciated by considering a specific example: with a partition coefficient of 50 : 1 and a volume ratio of 1 : 2, a single-stage extraction loses 3·8% of the input in the aqueous phase, against only 0·14% loss incurred during the two-stage process.

4.3. *Ion exchange resins and columns*

Products of ionic character when present in aqueous filtrates of fermentation broth may sometimes be concentrated and purified by ion exchange chromatography. Ion exchange resins are insoluble branched polymers of high molecular weight, containing labile ions that are in equilibrium with the ions in the surrounding solution. Commercial resins, usually cross-linked in varying degree to confer the requisite degree of insolubility and to diminish the tendency for swelling by aggregation of water molecules, are conveniently available in the form of spherical beads. The labile ions of the resins are provided by groups that are ionized at suitable pH values: weakly acidic (e.g. carboxyl) and weakly basic (e.g. amino) groups will be largely dependent on pH for exhibition of their maximum exchange capacity, but strongly acidic (e.g. sulphonic acid) and strongly basic (e.g. quaternary ammonium) groups will be relatively independent of pH, being completely ionized at all practical pH values:

$$\text{(i)}\quad \underset{\text{(cation exchanger)}}{\text{Polymer-acidic group}} \underset{\text{acid}}{\overset{\text{alkali}}{\rightleftharpoons}} \text{polymer anion}^{\ominus} + \underset{\text{(exchangeable cation)}}{H^{\oplus}}$$

$$\text{(ii)}\quad \underset{\text{(anion exchanger)}}{\text{Polymer-basic group}} \underset{\text{alkali}}{\overset{\text{acid}}{\rightleftharpoons}} \text{polymer cation}^{\oplus} + \underset{\text{(exchangeable anion)}}{OH^{\ominus}}$$

The $H^{\oplus}$ or the $OH^{\ominus}$ forms of cationic and anionic exchangers, respectively, may have a tendency to attack labile products when undergoing extraction. The resins are therefore usually employed in their neutral salt forms, cation exchangers in the sodium form and anion exchangers in the chloride form. Conversion from one form to another is governed by the law of mass action, so that the behaviour of the resin can be predicted by consideration of the equation:

$$\frac{[H^{\oplus}_R] \times [Na^{\oplus}_A]}{[Na^{\oplus}_R] \times [H^{\oplus}_A]} = K$$

where $[H^{\oplus}_R]$ = concentration of labile hydrogen ion on the resin;

$[H^{\oplus}_A]$ = concentration of hydrogen ion in the aqueous phase surrounding the resin;

$[Na^{\oplus}_R]$ = concentration of labile sodium ion on the resin;

$[Na^{\oplus}_A]$ = concentration of sodium ion in the aqueous phase surrounding the resin.

This equation becomes more universally true if activities (effective concentrations) replace concentrations. Clearly, either an increase in the concentration of sodium ion or a decrease in concentration of hydrogen ion in the aqueous phase, due to increased alkalinity, will tend to increase the concentration of labile sodium ion relative to the concentration of labile hydrogen ion on a cation exchange resin (i.e. will tend to put the resin into the "sodium cycle" by preferential adsorption of sodium ions). Similar considerations are valid for conversion of anion exchangers from the hydroxyl form to the "chloride cycle".

Some commonly encountered resins are included in the tables below.

In the same way that the efficiency of extraction is increased when a two-stage solvent extraction system is employed, so a two-stage ion exchange system is better than a single-stage one. The most efficient arrangement, however, is a column in which the solution flows slowly downwards over the static beads of resin. Such an arrangement may be considered to be composed of many theoretical stages or layers that do not mix (Fig. 2.7).

CATION EXCHANGERS

Grade	Ionic behaviour	Type of polymer	Functional groups
IR 120[1]	Strongly acid	Polystyrene	SO_3H
Zeo-Karb 215[2]	Strongly acid	Phenolic	SO_3H
IRC 50[1]	Weakly acid	Acrylic	COOH
Duolite CS 100[3]	Weakly acid	Phenolic	COOH

[1] Manufactured by Rohm & Haas Co.
[2] Manufactured by the Permutit Co.
[3] Manufactured by Chemical Process Co.

ANION EXCHANGERS

Grade	Ionic behaviour	Type of polymer	Functional groups
IRA 400[1]	Strongly basic	Polystyrene	$\overset{\oplus}{N}Me_3$
IR 45[1]	Weakly basic	Polystyrene	Polyamine
De-Acidite G[2]	Weakly basic	Polystyrene	NEt_2

[1] Manufactured by Rohm & Haas Co.
[2] Manufactured by the Permutit Co.

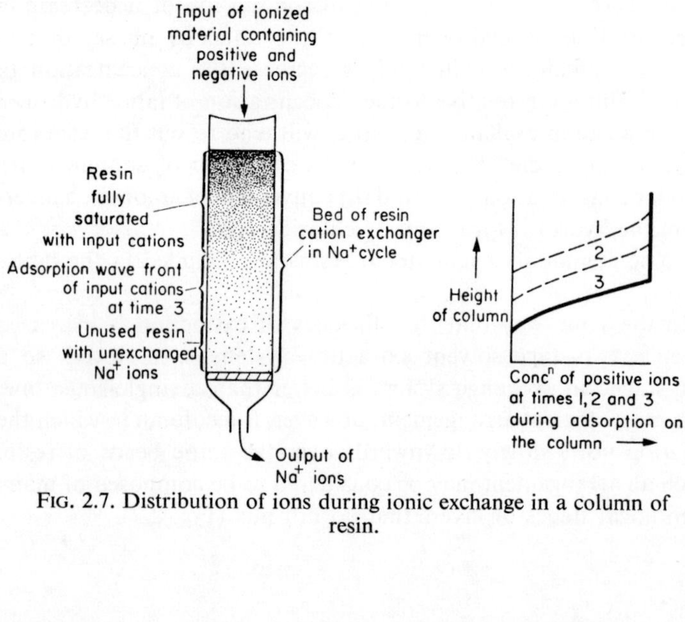

FIG. 2.7. Distribution of ions during ionic exchange in a column of resin.

The adsorption wave-front passes gradually down the column as adsorption of product from the input material proceeds. The shape of the wave-front determines the efficiency, and this in turn is determined by the kinetics of ion exchange and by the relative affinities for the resin of the ions, and is modified by practical considerations governing uniformity of packing and absence of channelling in the resin column. Ideally, the wave-front should be horizontal (i.e. abrupt). The rate of ion exchange depends on two diffusion coefficients: the coefficient of diffusion for the film surrounding the resin particles and the coefficient of diffusion within the particles. The thickness of the film surrounding the particles may be diminished by increased flow rate or agitation with concomitant increase in rate of adsorption. Increase in resin particle size naturally diminishes the rate of particle diffusion, but it also reduces film diffusion because of diminished turbulence. Increase in ionic concentration increases only the rate of film diffusion, but does not affect the rate of diffusion inside the resin particles.

The practical choice of resin type will depend upon the chemical and physical properties of both resin and product and on the nature of the contaminating substances: for example, chalk is sometimes omitted from fermentation media for streptomycin production because the resultant calcium ions would be adsorbed in preference to the streptomycin cations on the ion exchange resin used for extraction.

Extraction of streptomycin from broth filtrate by use of a carboxylic acid resin may be considered an example of an industrial extraction based on an ion exchange process. A series of two columns in the sodium cycle is used: the broth passes through the first until the resin is saturated with antibiotic cation (Fig. 2.7.), and then the second column becomes the first of a new series. The resin is prepared in the sodium form by passing sufficient 10% NaOH solution (20 × wt. of resin) slowly in an upward direction through the columns and washing with demineralized or distilled water to remove excess sodium ions. Filtered and clarified broth filtrate from a *Streptomyces griseus* fermentation is passed at the rate of 10 bed volumes per hour downward through the two columns in series to give almost complete adsorption. The flow rate in this example is remarkably high compared with those of other extractions in which the equilibrium conditions are less satisfactory. The resin is capable of adsorbing its own weight of streptomycin.

The resin is washed with demineralized water to remove excess broth, and the adsorbed streptomycin cations (streptomycin $H_{\oplus}$) are then eluted by passing 0·1 N HCl (1% broth volume) down at a slow flow rate through the individual columns, thereby converting the resin to the hydrogen form and eluting the streptomycin as the hydrochloride. Thus in one step the antibiotic has been both purified and one hundred times concentrated.

4.4. *Roller dryers*

When the required product is present in the mycelium or insoluble residue, purification may be effected by stirring with a suitable solvent, filtering and evaporating the extract. An intermediate step useful for stable non-volatile products is roller drying either the whole broth or the solid residue. The liquid to be dried is fed to a reservoir formed by the space between the two slowly rotating (1–2 rev/min) steam-heated (up to 100 psig) drums (Fig. 2.8). The gap between the drums, termed the "nip", is

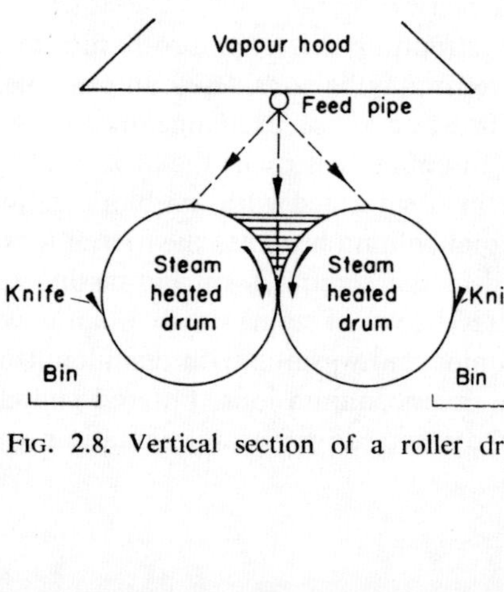

FIG. 2.8. Vertical section of a roller dryer.

adjusted to permit uniform flow of liquid into a thin film on the surface of each drum. Rapid evaporation of water deposits solids on the surface of each drum, and these solids are removed by scraping with fixed knife-blades. Adjustment of the gap depends on the physical properties of the material to be dried: if the gap is too wide, the product will remain wet (probably sticky), but if the gap is too narrow, the throughput will be inadequate.

The advantage of roller drying lies in provision of a free-flowing powder with minimum "cooking". The process may be conveniently used for production of animal-feed material from riboflavin, tetracycline or vitamin B_{12} fermentation broths; as an intermediate stage in an extraction process, e.g. griseofulvin; and for the drying of milk, cereal products or yeast. The disadvantages of roller drying include the relatively high capital, maintenance, and operating costs.

4.5. *Flame-proof areas*

When solvents are used for extraction of the product, provision must be made for mixing the broth with a solvent. Many solvents employed in industry are inflammable, and they are therefore handled in a flame-proof area. The fire hazard of a solvent depends upon its "flash point". This is defined as the temperature above which the vapour forms in air an ignitable mixture above the solvent: this mixture will ignite only when triggered by material already at or above the "ignition temperature".

An examination of the properties of various common solvents listed in Table 2.2 warns the research worker and designer that carbon disulphide is best avoided and in any event should only be used in closed vessels, that ethyl ether should be replaced by isopropyl ether whenever possible, and that all other solvents should be handled with care in a flame-proof area.

Electric motors, light fittings and switches used in flame-proof areas must be sealed, to prevent sparks from igniting inflammable vapours. Static electricity is eliminated by thorough earthing of all equipment, especially any movable metal containers. Care must

TABLE 2.2. PROPERTIES OF SOME COMMON INDUSTRIAL SOLVENTS

Solvent	Ignition temperature (°C)	Flash point (°C)
Acetone	561	—18
Benzene	580	—11
Butyl acetate	421	29
Butyl alcohol	345	38
Carbon disulphide	120	—30
Cellosolve	238	40
Ethyl acetate	484	— 2
Ethyl alcohol	392	12
Diethyl ether	193	—45
Methanol	470	11
Light petroleum	329	—56
Isopropyl ether	443	—28

also be taken to prevent development of high potentials when liquids move over the surfaces of plastic apparatus. Smoking is prohibited, the carrying of matches or of cigarettes or any other form of tobacco is forbidden, and workers entering the flame-proof area are made to wear non-spark safety shoes (fabricated from conductivity rubber), and are trained to use only the traditional "rubber hammer" or other flame-proof tools.

4.6. *Other equipment*

Additional equipment needed for a fermentation unit will depend on the extraction processes used. Flame-proof "vacuum" distillation is necessary to permit concentration of solvents at low temperatures (20–50°C); distillation columns are needed for purification and recovery of the solvents used in extraction processes; lyophilization or freeze-drying apparatus may be required to remove water as a sublimate from ice within the frozen product to give a finely divided and free-flowing powder, as in the final stages of streptomycin sulphate production; large-scale laboratory-type filters are convenient for filtration of intermediate grade products precipitated from aqueous or non-aqueous solution; plate-frame filters fitted with asbestos pads permit both sterilization and removal of pyrogens (polysaccharide substances that cause a rise in body temperature when injected) from solutions of products before freeze-drying or precipitation; charcoal columns may be used for adsorption or for decolorization of products; and so on. The products from fermentations are diverse and the extraction method employed must be devised for the individual product: economics will arbitrate between alternative processes.

Some manufacturers of extraction equipment

1. *Centrifugal separators*
   Podbielniak Inc., 341 E. Ohio Street, Chicago, Ill., U.S.A.
   Alpha Laval Ltd., Great West Road, Brentford, Middlesex, England.
   Sharples Centrifuge Ltd., "Tower House", Woodchester, Stroud, Glos., England.

2. *Filters*
   Filtration Engineering Inc., Newark, N.J., U.S.A.
   British Filters Ltd., Old Court, Cox Green, Maidenhead, Berks., England.
   Carlson-Ford Sales Ltd., Newman Street, Ashton-under-Lyme, Lancs., England.
   Doulton Industrial Porcelains Ltd., Royal Doulton Potteries, Wilnecote, Tamworth, Staffs., England.
   Stockdale Engineering Ltd., Filtration Division, London Road, Poynton, Ches., England.

Suggested literature for further reading

1. CALMAN, C. and KRESSMAN, T. R. E., *Ion Exchangers in Organic and Biochemistry*, Interscience, New York, 1957.
2. STEEL, R. (ed.), *Biochemical Engineering*, Heywood, London, 1958.

CHAPTER 3

Industrial Applications of the Genetics of Micro-organisms

THE industrial utility of any particular strain of micro-organism is determined, to a large extent, by a characteristic combination of hereditary factors. Most of the hereditary information in living organisms is in the chromosomes of cell nuclei and is transmitted to succeeding generations by the agency of an undetermined number of molecules of deoxyribonucleic acid (DNA), occurring as a complex with basic proteins, in each of these structures.

1. Structure of deoxyribonucleic acid (DNA)

Each molecule of DNA comprises two linear polymers of nucleotides (Fig. 3.1) arranged as a double helix. The deoxyribose sugar residues and the phosphate groups of the individual nucleotides provide the external form to this spiral structure, but the organic nitrogenous bases—adenine, guanine, cytosine and thymine (methyl uracil)—project into the interior so that pairs of bases, one from each polymer chain, meet at right-angles to the main molecular axis in a series of spirally displaced planes. The junction between each two bases is effected by hydrogen bonding, and this determines that adenine must pair with thymine and guanine with cytosine (Fig. 3.2).

2. The genetic code

It has been found that within the DNA chains of the chromosomes the precise sequence of bases, probably taken in triplets,

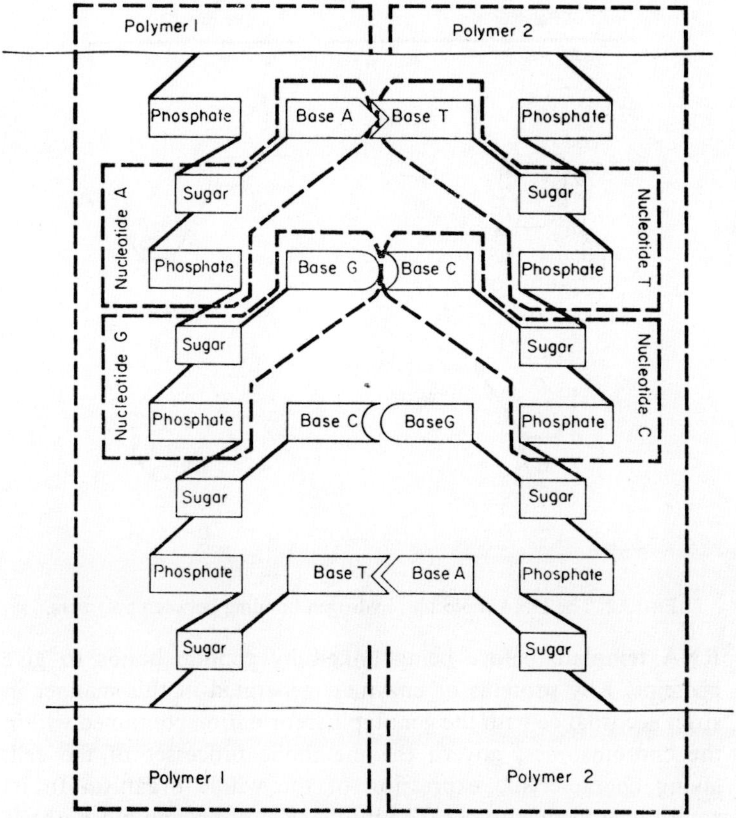

FIG. 3.1. The DNA molecule: diagrammatic structure.

determines the genetic constitution (genotype) of the cell by providing a specific and ordered sequence of genes (factors determining hereditary characteristics) along the length of each chromosome. This genotypic information is carried by "messenger ribonucleic acid" (mRNA) to the ribonucleoprotein particles (ribosomes) of the cytoplasm. Meanwhile, amino acids, activated by adenosine triphosphate (see Chapter 9), are moved to the same site by specific "soluble" or "transfer ribonucleic acids" (sRNA), where they are assembled in ordered sequence on the messenger

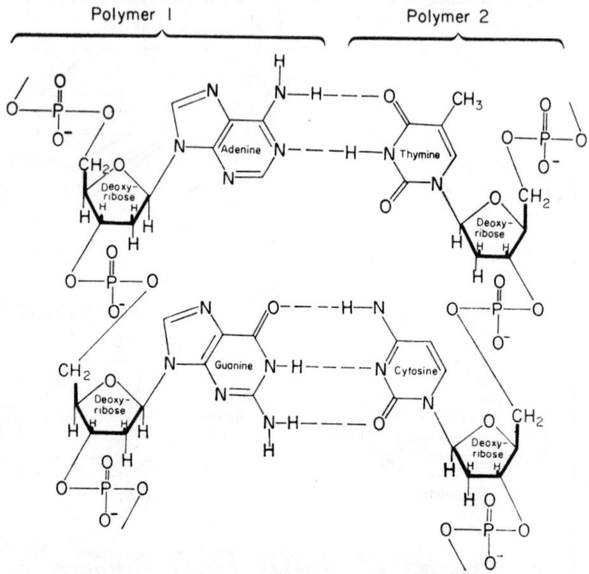

FIG. 3.2. The DNA molecule: hydrogen bonding between polymers.

RNA template before being linked by peptide bonds to give proteins. Key proteins or enzymes, generated in this manner in strict accordance with the genotypic information contained within the chromosomes, govern the metabolic processes of the cell, giving characteristic expression of the whole organism to its environment (phenotype). The mechanism of protein biosynthesis and the formation of cellular enzymes will be given more detailed consideration in Chapter 8.

The molecules of DNA are also capable of self-duplication, each of the two constituent polymer strands functioning as a template for a new complementary strand. The mechanism of this process is not yet understood, but it is probably the basis for the observed duplication of chromosomes that precedes transfer (aided by the achromatin fibres of the "spindle mechanism") of one chromosome from each duplex to each of two daughter nuclei during vegetative cell division (mitosis).

3. The chromosome theory

The number of chromosomes (chromosome complement) is constant for a species. The life-cycle of organisms takes many forms, but those capable of sexual reproduction exhibit an alternation of generations in which a phase of growth having nuclei with the basic (haploid) number of chromosomes alternates with a form having a double (diploid) set of chromosomes in its nuclei. Replication of the diploid set of chromosomes at each mitotic division is interrupted at some stage in the life-cycle by a special form of cell division in which the chromosome complement is halved by an initial pairing of homologous chromosomes (chromosomes with the same sequence of genes) before one from each pair is transferred to each of two haploid daughter nuclei: this is meiosis. Reconstitution of the diploid phase occurs when nuclei from parent haploid forms fuse their chromosome complements to give an initial diploid nucleus or fertilized egg.

The relative importance in the life-cycle of the haploid and diploid phases may differ widely in different organisms: in many fungi and in some actinomycetes an intermediate phase has been termed a heterokaryon, because genetically distinct haploid nuclei reproduce separately within the same cell.

The pairs of homologous chromosomes in the diploid complement are frequently not identical. This arises because certain portions of the genotypic strands can occur in different forms: when these can be recognized by genetic analysis they are termed alleles. Alleles are different forms of the same gene that have arisen in the course of time by spontaneous mutational change occurring with a low frequency. The mutation rate may be artificially increased by the methods described in the next section. Mutations may afford many alleles of the same gene, but they must occupy homologous loci (corresponding positions in a pair of homologous chromosomes), so that only one allelic pair can occur in each diploid nucleus. The alleles are reproduced at each cell division when the chromosomes themselves are replicated.

Most inherited characteristics can be related to differences between alleles in homologous chromosomes. Pairs of characters determined by genes within homologous chromosomes show linked inheritance, but those governed by genes in non-homologous chromosomes show independent inheritance. Exchanges between similar portions of homologous chromosomes are termed "genetic cross-overs": they can occur during meiosis, and this provides a mechanism for limited recombination between linked genes. In bacteria the linearly arranged genes appear to form a single linkage group.

4. The parasexual cycle

The parasexual cycle replaces sexual reproduction in imperfect fungi (fungi lacking the sexual spore stage) and in some actinomycetes, but affords similar possibilities for genetic recombination. In this process the genetically different haploid nuclei of a vegetative heterokaryon (a cell, or more usually, mycelium, containing more than one genetically distinct type of nucleus in a common cytoplasm) occasionally undergo fusion to give a diploid heterozygous nucleus that can multiply by mitosis along with other parent haploid nuclei that remain in the heterokaryotic condition.

Occasionally during multiplication the heterozygous diploid nuclei give nuclei that, although still diploid, are now homozygous with respect to certain loci that were originally heterozygous. The explanation of this phenomenon is attributed to "mitotic crossing-over", a process of exchange at mitosis, between similar portions of homologous chromosomes, that is probably not basically different in mechanism from the meiotic crossing-over already considered. Suitable microbiological techniques permit establishment of these homokaryotic diploids and some may exhibit a desirable combination of genetic characters.

Finally, haploidization of a small proportion of all diploid nuclei occurs, and this results in random re-assortment of whole chromosomes: many of these haploid forms will necessarily be eliminated because their genetically unbalanced constitution renders them non-functional.

The importance of the parasexual cycle will depend upon the frequency of the individual stages in the sequence, but in some fungi of industrial importance, e.g. *Penicillium chrysogenum* and *Aspergillus niger*, the rate is sufficiently high to be an important source of genetic variation.

5. Types of mutation and some mutagenic agents

The genetic recombinations occurring at meiosis and during the somatic crossing-over of the parasexual cycle result in small changes of genotype, and these provide a minor source of genetic variation. Abrupt and major change of genotype has greater industrial utility, and this is generally due to mutation, whereby the character of one allele or of a pair of alleles has been altered. Most industrially important mutations of micro-organisms result from changes within nuclei of three types.

(i) Change of chromosome number (by addition to or subtraction from a diploid cell of a whole chromosome, or more than one, or even of a haploid chromosome set).
(ii) Structural changes that alter the number or the arrangement of alleles within chromosomes (by chromosome breakage resulting in inversion of a segment of the chromosome or deletion of a portion of it).
(iii) Qualitative changes in the genes themselves, resulting in emergence of new alleles: these are gene mutations.

It is usually difficult to distinguish the nature of the change but in industry this is fortunately not important. It is only necessary to understand that chromosomes are susceptible to mutation in several different ways, and therefore that mutagenic agents, operating by different modes of action, may be expected to induce different changes in genotype. Some of these may increase the potential yield of the original product or may result in production of new structurally related substances.

Artificial mutations that are due to change of chromosome number usually result from the action of chemicals that inhibit functioning of the "spindle mechanism" during mitotic cell division. On the other hand, mutations induced by chromosome breakages may be effected by action of the mutagen on either DNA itself or on the protein associated with the DNA of the chromosomes, and the ultimate effect will depend upon the distribution of the fragments between the daughter cells.

Gene mutations are to be expected when physical or chemical energy derived from the mutagenic agent effects a change in molecular structure of DNA at the gene sites or when the genes are prevented from normal replication by inhibition of DNA biosynthesis.

5.1. *Producing and isolating mutants*

Induction of mutations in micro-organisms must, of necessity, be done on a population, and it is usual to employ a population termed a "clone" that has been propagated by mitosis from a single cell, monospore or hyphal tip. The result of mutagenic treatment is expressed in terms of the frequency of mutations induced in the original population. Segregation of mutant characters in ratios according with experience of Mendelian inheritance depends upon occurrence of genetic cross-overs during meiosis: observance of these ratios is the best proof of the hereditary and nuclear origin of the mutation. Analysis of mutations in micro-organisms having a parasexual cycle may be done in a similar manner, but the expected ratios will be different. In such cases an aberrant strain is considered to be a mutant if it shows all-or-nothing segregation for the presumed mutant phenotype when it is made into a diploid by growth in association with the normal strain.

It is therefore necessary to devise screening methods that permit none but mutant cells to be counted.

(i) The original population may be grown on the simplest medium that will support growth ("minimal medium"): nutritional mutants resulting in loss of ability to synthesize a particular metabolite, known as "auxotrophs", may then be detected by the fact that they will grow only when the mutationally inhibited factor is added to the medium. Modification of this method permits detection of reverse mutations in an auxotrophic strain, known as "prototrophs", in which the original biochemical genotype is restored.

(ii) Antibiotics or poisons that inhibit the growth of the original population may be incorporated in the medium, with the result that mutants resistant to these substances may be readily detected. Bacteriophages may be employed in similar manner to detect phage-resistant mutants.

(iii) The giant colony technique is sometimes useful for mutations that can be recognized visually. For this purpose, a petri dish containing agar medium with mutagenic agent incorporated (e.g. 0·2 mC ^{35}S as sulphate in each 20 ml) is inoculated at the centre with the experimental culture. On incubation, radial growth occurs and mutations may be recognized as sectors having different morphological appearance, colour, degree of sporulation or rate of growth. A minimum amount of inoculum is taken from the perimeter of any given mutant sector for propagation of new colonies on medium free from mutagenic agent.

When the giant colony technique is employed, homogeneity of the mutated strain must finally be established by isolation of monospores in the usual way. It is also important to remember that changes in colour or in morphology may only be temporary modifications resulting from cytoplasmic, as opposed to nuclear, changes. These characters can only be used to detect nuclear mutants when prior information exists to indicate that the two are associated.

5.2. *Mutagenic agents*

Most mutagenic agents are toxic, and the maximum mutation frequency is generally observed when the dosage permits only a small number of survivors (frequently of the order of 0·1%): the actual dose to be employed in practice is determined experimentally.

A variety of mutagenic agents that have value or potential use in industry for production of mutant micro-organisms may be grouped in categories according to their three probable modes of action, but it is possible that some operate by more than one mechanism.

(i) Mutagenic agents causing change of chromosome number include colchicine, isopropylphenylcarbamate, trichloroacetanilide and the γ-isomer of hexachlorocyclohexane.

(ii) Mutagenic agents causing structural changes in the number or the arrangement of alleles within chromosomes include X-rays, ultraviolet light, nitrogen mustards, a variety of radioisotopes (especially ^{35}S), foreign DNA and the metal-chelating agent, ethylene-diamine-tetracetic acid (EDTA).

(iii) Mutagenic agents producing qualitative changes in the genes themselves include chemicals inhibiting enzymes concerned in biosynthesis of DNA, competitive analogues of purine and pyrimidine bases, nitrous acid, hydroxylamine and the acridines.

It may be that, as more is learned of industrial micro-organisms exhibiting a parasexual cycle, practical use may be made of the already demonstrated possibilities of parasexual breeding as a source of genetic variation.

Suggested literature for further reading

1. Alikhanian, S. I., Induced mutagenesis in the selection of micro-organisms, *Adv. appl. Microbiol.*, **4**, 1–50 (1962).
2. Auerbach, C., *Mutation, An Introduction to Research on Mutagenesis, Part 1. Methods*, Oliver & Boyd, London, 1962.
3. Bennett, J. C. and Dreyer, W. J., Genetic coding for protein structure, *A. Rev. Biochem.* 205–34 (1964).
4. Bradley, S. G., Parasexual phenomena in micro-organisms, *A. Rev. Microbiol.*, **16**, 35–52 (1962).

5. CATCHESIDE, D. G., *The Genetics of Micro-organisms*, Sir Isaac Pitman, London, 1951.
6. CAVALIERI, L. F. and ROSENBERG, B. H., Nucleic acids and information transfer, *Prog. Nucleic Acid Research* **2,** 1–18 (1963).
7. CRICK, F. H. C., BARNETT, L., BRENNER, S. and WATTS-TOBIN, R. J., General nature of the genetic code for proteins, *Nature, Lond.* **192,** 1227–32 (1961).
8. CRICK, F. H. C., GRIFFITH, J. S. and ORGEL, L. E., Codes without commas, *Proc. natn. Acad. Sci. U.S.A.* **43,** 416–21 (1957).
9. CRICK, F. H. C., The biological replication of macromolecules, *Symp. Soc. exp. Biol.*, No. 12, 138–63 (1958).
10. CRICK, F. H. C., The recent excitement in the coding problem, *Prog. Nucleic Acid Research* **1,** 163–217 (1963).
11. DE DEEPESH, N., A new chromosome model, *Nature, Lond.* **203,** 343–6 (1964).
12. FINCHAM, J. R. S. and DAY, P. R., *Fungal Genetics* (Botanical Monographs **4**), Blackwell, Oxford, 1963.
13. GALE, E. F., *The Synthesis of Proteins and Nucleic Acids, The Bacteria.* III: *Biosynthesis*, 471–576, Academic Press, New York and London, 1962.
14. HAYES, W. and CLOWES, R. (eds.), *Microbial Genetics* (*10th Symp. Soc. Gen. Microbiol.*), Cambridge University Press, 1960.
15. HURWITZ, J. and AUGUST, J. T., The role of DNA in RNA synthesis, *Prog. Nucleic Acid Research* **1,** 59–92 (1963).
16. JACOB, F. and MONOD, J., Genetic regulatory mechanisms in the synthesis of proteins, *J. molec. Biol.* **3,** 318–56 (1961).
17. KIT, S., Deoxyribonucleic acids, *A. Rev. Biochem.* **32,** 43–82 (1963).
18. KRIEG, D. R., Specificity of chemical mutagenesis, *Prog. Nucleic Acid Research* **2,** 125–68 (1963).
19. LIPMANN, F., Messenger ribonucleic acid, *Prog. Nucleic Acid Research* **1,** 135–61 (1963).
20. NEIDHARDT, F. C., The regulation of RNA synthesis in bacteria, *Prog. Nucleic Acid Research* **3,** 145–81 (1964).
21. PARDEE, A. B., Gene expression: its specificity and regulation, *A. Rev. Microbiol.* **16,** 1–34 (1962).
22. PONTECORVO, G., The parasexual cycle in fungi, *A. Rev. Microbiol.* **10,** 393–400 (1956).
23. SINSHEIMER, R. D., Is the nucleic acid message in a two-symbol code?, *J. molec. Biol.* **1,** 218–20 (1959).
24. STEVENS, AUDREY, Ribonucleic acids—biosynthesis and degradation, *A. Rev. Biochem.* **32,** 15–42 (1963).
25. UHL, C. H., A model of sister-strand crossing over, *Nature, Lond.* **206,** 1003–5 (1965).
26. WARREN, L., *Nucleotides and Nucleosides. Metabolic Pathways II*, pp. 459–524. Academic Press, New York and London, 1961.
27. WEISS, J. J., Chemical effects of ionizing radiations on nucleic acids and related compounds, *Prog. Nucleic Acid Research* **3,** 103–44 (1964).

CHAPTER 4

Culture Collection Laboratory

THE potential for production of industrial fermentation products resides wholly in the particular strain of micro-organism employed: environmental conditions can markedly affect both end-product and yield, but only as far as the genetic constitution of the strain permits. It is therefore vitally important to preserve the homogeneity of industrial micro-organisms by maintenance under conditions that retain viability but prevent loss of strain characteristics: this is usually a function of the culture collection laboratory.

The culture collection laboratory should preferably contain a small room supplied with sterile air and held under slight positive pressure for use during sub-culturing and inoculation. Sterile transfers can also be done in the laboratory if precautions are taken to maintain still air by closing windows and doors. Periodic washing-down of walls, ceiling, benches, light fittings and furniture, accompanied by periodic disinfection with aerosols, helps to reduce contamination with spores. Potential for contamination is greatly increased under certain climatic conditions (e.g. spring-time in the United Kingdom) and when building excavations are being made near the laboratory.

1. Preservation of micro-organisms

Micro-organisms known to produce specified fermentation products are frequently listed in the catalogues of national culture collections: sometimes these strains have been deposited to comply with conditions for a grant of patent rights, but it would be unwise to assume that the deposited strains reflect the yield potential of those currently employed by the patentee for commercial production. Alternatively, useful industrial strains may sometimes be found in soils and other natural environments by adaptation of the screening methods to be described in Chapter 18.

Whatever their origin, the potential for yield of the required product is subsequently increased by mutation and selection of improved strains in the manner described in Chapter 3. These possible production strains should be homogeneous and free from infection.

1.1. *Techniques for homogeneous cultures*

Homogeneity can be ensured by culturing a colony from a monospore that has been selected by micromanipulation: when this is not feasible, a colony formed from the apex of a hypha isolated in similar manner is usually satisfactory. Freedom from general infection is aided by application of the principles of asepsis described in Chapter 5. Nevertheless, occasional contamination may occur, and its recognition at an early stage is important. Complete records for each culture, noting the rate of growth and morphological appearance at each stage, and the microscopical appearance of living and stained cultures afford invaluable information: this should be recorded at each sub-culture, together with the results of yield tests.

Contamination by a species from a different genus is sometimes readily detected by cursory inspection: plating out the less heavily infected portion of the culture may then permit selection of fresh uncontaminated typical colonies. Infection by morphologically indistinguishable species is more difficult, because their presence is usually only evident from decreased yields: reselection by yield tests of new monospore or hyphal-tip isolates may then be the only remedy.

Contamination in a collection of diverse micro-organisms is minimized when cultures are stored and sub-cultured in random order: chance infection may then be expected to be more readily detected because of probable differences in morphology.

Antibiotics, because of their properties of differential toxicity, have also proved useful when incorporated in maintenance media for prevention of chance contamination and even as an aid to decontamination. Cycloheximide (2–20 μg/ml) inhibits the growth of yeasts in cultures of bacteria and some fungi and is useful in 100 μg/ml concentration for inhibiting fungi in cultures of bacteria or actinomycetes; neomycin sulphate (20 μg/ml) limits contamination of fungal cultures by actinomycetes; the sodium salt of benzylpenicillin (40 μg/ml) protects cultures of fungi and of actinomycetes against many bacterial infections; griseofulvin (2–20 μg/ml) is sometimes useful for its selective antifungal action; and a mixture of oxytetracycline hydrochloride (5 μg/ml) with the polyene antibiotic lagosin (20 μg/ml) will protect cultures of actinomycetes against many fungal and bacterial contaminants.

Mites, by carrying spores throughout the culture collection, are one of the most frequent causes of contamination: cotton-wool plugs or petri dish lids are no barrier. Mites feed on the cultures, and they may usually be detected as small light-coloured objects if set in motion by gently warming the suspect culture: they may be readily observed, with their eggs of similar size, when viewed under a dissecting microscope.

Infestation can occur even under refrigerated conditions, but some protection is afforded by standing in water the legs of any cabinet where cultures are stored. Vapour generated at room temperature by crystals of *para*-dichlorobenzene contained in a sealed chamber is sometimes effective in killing mites in contaminated cultures, but the chemical may be toxic to some species of fungi. 1,1-Bis-(chlorophenyl)-trichloroethanol (available under the proprietary name of Kelthane as a concentrate for atomizing or as an emulsion for dilution with water) is to be preferred, because it is lethal to all stages in the life-cycle of mites. Kelthane-containing fluid may also be used for routine cleansing of laboratory benches and store shelves. Mite-infested cultures can be freed from contamination by sub-culturing two or more times, as necessary, on agar medium appropriate for the mould concerned and supplemented with Kelthane (0·025% of active ingredient).

1.2. *Instability of cultures*

In order to maintain viability, homogeneous stock cultures must be periodically sub-cultured on appropriate media and at optimum temperatures (usually 25°C for fungi, 28°C for actinomycetes, 37°C for bacteria, and 60°C for thermophilic micro-organisms), at intervals of 1–6 months, depending on the characteristics of the particular organism. This procedure introduces fresh problems. Some moulds replicate themselves in successive sub-cultures under artificial conditions, but others show varying degrees of instability expressed in loss of ability to form conidia or fruiting structures, in loss of growth vigour, in "sectoring", or in loss of some biochemical characteristic, such as the ability to produce a particular metabolite.

The tendency to instability in sub-cultures can be minimized by attention to environmental conditions. Loss of ability to sporulate or to produce fruiting bodies may sometimes be countered by exposure to diffuse light, by brief exposure to sunlight, by sub-culture on media containing a low concentration of available nitrogen, or sometimes by use of media with a low concentration of available carbohydrate. Some fungi recover their ability to sporulate when provided with accessory food factors, and a few respond only to the high osmotic pressure provided by 20–40% w/v sucrose. It is notable that anhydroglucose has recently been shown to be the trace ingredient in glucose responsible for stimulating conidium production by *Penicillium* spp., and this chemical may find a useful role in the culture collection laboratory. In submerged cultures of some moulds, sporulation is facilitated by interaction of high aeration with low concentration of nitrogen: occasionally, sporulation in submerged culture may be induced by inclusion of $CaCl_2, 6H_2O$ (2·5 w/v) in the medium (e.g. in *Penicillium chrysogenum*).

Because of this tendency to instability in sub-cultures, micro-organisms are most conveniently stored over long periods in a metabolically inhibited state and preferably as replicated cultures to insure against accidental loss. Refrigeration of cultures at 4°C is helpful, but some moulds are able to continue growth at this temperature. Flooding with pure sterile mineral oil facilitates storage of some tube cultures and is particularly useful for basidiomycetes. Lyophilization or freeze-drying is to be preferred for cultures producing abundant conidia or other types of spores resistant to desiccation: measured portions of the sporulated cultures are lyophilized in ampoules and held over a desiccant until sealed under dry nitrogen. Actinomycete cultures, in particular, can be stored for many years under these conditions without change of character.

The constitution of the medium for suspension of spores largely governs successful preservation by lyophilization: when difficulties are experienced they may be overcome, in the last resort, by adding the spore suspension, in an amount just adequate for saturation, to sterile dry, finely sieved, dark-coloured loam soil, having a natural pH of 6·6–7·0, before freeze-drying.

2. Preparation of inoculum

The culture collection laboratory also provides inoculum for initiating fermentations in fermenters of many different sizes. Inoculum for laboratory submerged culture fermentations might typically be prepared as described below. Freeze-dried spores of the required strain are re-suspended by addition of sterile distilled water to the ampoule. Sufficient spore suspension to form a film is transferred to slopes of agar maintenance medium in $3\frac{1}{4} \times 1$ in. cylindrical glass bottles fitted with screw caps permitting adequate aeration when closed, but not screwed tight (small surface culture bottles). After incubation at the appropriate temperature and for the time needed to give well-grown and sporulated cultures, 5–10 ml sterile water is added to each culture, and the surface is scraped before transferring 2 ml of the suspension of spores and mycelium as a film to the surface of maintenance agar medium in flat 10 oz medicinal bottles plugged with cotton-wool to permit adequate aeration (standard surface culture bottles). When particulate media are used for this type of culture it is important to ensure uniform distribution of nutrient ingredients throughout the agar, so that the mycelium developing on incubation sporulates evenly over the entire surface of the colony. Complete sporulation may need about 7–14 days, depending on strain and temperature. The cultures may be stored (sporulated surface downwards, to prevent condensate settling on the culture) at 4°C for up to one month before being used as initial source of inoculum for laboratory submerged fermentations aerated by diffusion, for mechanically agitated and aerated laboratory fermenters, or for plant-type fermenters.

Inoculum for pilot or production plants is usually despatched as a submerged culture. For this purpose 10 ml sterile water is added to the sporulated culture in a standard surface culture bottle before scraping the surface growth to give a suspension of spores and mycelium. Of this inoculum 5–10% v/v is transferred to 2 l. conical or round flasks containing about 400 ml inoculum development medium, before the liquid cultures are incubated under conditions suitable for submerged culture of the particular organism. The prepared inoculum may be required as vegetative mycelium, in either the logarithmic or the retardation phase of growth (see Chapter 8), or as a fully sporulated culture; the requisite environmental conditions of medium, temperature and aeration need to be determined by experience for each strain. In general, vegetative inoculum can usually be obtained in 2–3 days, but submerged spores may need 7–14 days. Sterile transfer is made of the submerged culture inoculum to an aspirator or other container that may be used conveniently for inoculation of the seed stage of pilot plant or production plant fermentation broths. A residual aliquot of the submerged culture inoculum is retained for a routine check of freedom from contamination.

3. Yield tests in shake-flask fermentations

All preserved strains must be checked periodically to determine that no loss of ability to produce the desired end-product has occurred. This is done by conducting, under standard and predetermined optimum conditions, replicated fermentations in shake-flasks (250 ml or 100 ml conical flasks plugged with cotton-wool and containing 25–60 ml or 10–24 ml, respectively, of medium aerated by agitation on a rotary shaker). The procedure for a typical shake-flask fermentation is summarized below:

Sporulated surface culture
(standard or small
surface culture bottle)

↓ 5–10% v/v spore/mycelium suspension

Shake-flasks for
production of
submerged culture
(inoculum development
medium)

↓ 5–10% v/v submerged vegetative mycelium

Shake-flasks for
submerged fermentation
(fermentation medium)

The fermentation broths are assayed for concentration of the required product by aseptic removal of measured portions at suitable intervals of time.

It is usual to make yield tests when changing from one batch of inoculum ampoules to another, from one soil culture to another, or when a particular strain has not been used for an extended period of time. Similarly, in production plants it is common practice for the culture collection laboratory to make yield tests of a portion of each preparation of submerged culture inoculum supplied for initiation of each fermentation. Techniques for yield testing also find application in research laboratories when environmental conditions are deliberately altered to determine the effects of experimental conditions or strains in the standard fermentation.

4. National type-culture collections

Commonwealth Mycological Institute, Collection of Fungus Cultures, Ferry Lane, Kew, Surrey, England.

Culture Collection of Algae and Protozoa, Botany School, Cambridge, England.

National Collection of Type-Cultures, Central Public Health Laboratory, Colindale Avenue, London, N.W.9., England.

National Collection of Yeast Cultures, Brewing Industry Research Foundation, Lyttel Hall, Nutfield, Nr. Redhill, Surrey, England.

Institut Pasteur, Paris, France.

Institute for Fermentation, Microbiological Type-Culture Collection, 4–54, Juso-Nishinocho, Higashiyodogawa-Ku, Osaka, Japan.

Centraalbureau voor Schimmelcultures, Javalaan 20, Baarn, Netherlands.

The National Collection of Industrial Bacteria, Department of Scientific and Industrial Research, Torry Research Station, P.O. Box 31, 135 Abbey Road, Aberdeen, Scotland.

Centre de Collections de Types Microbiens, 19 Rue César-Roux, Lausanne, Switzerland.

American Type-Culture Collection, 12301 Parklawn Drive, Rockville, Maryland 20852, U.S.A.

Culture Collection Unit, Fermentation Division, Northern Utilization Research Branch, U.S. Dept. of Agriculture, Peoria 5, Illinois, U.S.A.

U.S.S.R. Antibiotics Research Institute, Moscow, U.S.S.R.

Suggested literature for further reading

1. ARMSTRONG, J. J., ENGLAND, D. J. F., MORTON, A. G. and WEBB, J. A., Stimulation of sporulation in *Penicillium* by anhydroglucose, *Nature, Lond.* **197,** 723 (1963).
2. CALAM, C. T., The selection, improvement and preservation of micro-organisms, *Prog. ind. Microbiol.* **5,** 1–54 (1964).
3. CHATIGNY, M. A., Protection against infection in the microbiological lab: devices and procedures, *Adv. appl. Microbiol.* **3,** 131–92 (1961).
4. GREEN, S. R., *A Review of Differential Techniques in Brewing Microbiology, Wallerstein Advances in Beer Quality*, Part II, Wallerstein Laboratories Communications **25** (87) 246–55 (1962).
5. HECKLY, R. J., Preservation of bacteria by lyophilization, *Adv. appl. Microbiol.* **3,** 1–76 (1961).
6. MILLER, M. W., *The Pfizer handbook of Microbial Metabolites*, McGraw-Hill, New York, Toronto and London, 1961.
7. MUGGLETON, P. W., The preservation of cultures, *Prog. ind. Microbiol.* **4,** 190–214 (1962).
8. SHIBATA, S. and NATORI, S. (Faculty of Pharmaceutical Sciences, University of Tokyo) and UDAGAWA, S. (National Institute of Hygienic Sciences, Tokyo), *List of Fungal Products*, 1961.
9. SMITH, G., *An Introduction to Industrial Mycology*, 5th edn., Edward Arnold, London, 1960.
10. SOKOLSKI, W. T. and CARPENTER, O. S., Microbiological assay, *Prog. ind. Microbiol.* **1,** 95–135 (1959).
11. WAKSMAN, S. A. and LECHEVALIER, H. A., *Actinomycetes and their Antibiotics*, Williams & Wilkins, Baltimore, U.S.A., 1953.
12. WAKSMAN, S. A., *The Actinomycetes III: Antibiotics of Actinomycetes*, Baillière, Tindall & Cox, London, 1962.

CHAPTER 5

Principles of Sterilization, Sterility Tests and Asepsis

A STERILE medium is free from living organisms, but "sterile fermentation" denotes, by common usage, a fermentation broth in which only the desired fermenting organism can be detected. Although the sterile state is absolute, methods of testing for sterility are relative. Thus an infecting organism may be present in fermentation broth, but may remain undetected because the sterility testing medium is unable to support the growth of this particular organism.

A contaminated fermentation broth suffers in a variety of ways.

(i) Available nutrients are converted to unwanted products.

(ii) Environmental conditions are changed, so that the fermenting organism is unable to produce, or the medium is unable to maintain, the desired product. For example, an alkali-producing contaminant in a penicillin fermentation broth will cause the pH to rise sharply, thus destroying the penicillin already formed and preventing the *Penicillium chrysogenum* from producing a further quantity of penicillin.

(iii) Enzymes that destroy the product may be formed: for example, some bacteria rapidly decompose penicillin through the agency of penicillinase.

1. Techniques of sterilization

Techniques used for sterilization fall into six groups; each has its part to play in industry.

(i) High pressure steam.
(ii) Pasteurization and tyndallization.
(iii) Dry heat.
(iv) Chemicals.
(v) Radiation.
(vi) Filtration.

1.1. *High pressure steam*

Most industrial fermentations need a sterile medium in the fermenter at the start. The requirement is fulfilled when the temperature of the medium is raised to 120°C by forcing high pressure live steam through every inlet into the fermenter: it is sometimes supplemented by passing steam through heating-coils on the internal walls of the fermenters. Sterilization of solid particles in the medium is assured if adequate agitation is provided during the sterilization cycle. Every part of the fermentation liquor and equipment must be held at sterilization temperature for at least 15 min. Cold pockets in which non-sterile fluid could collect, for example, in the sampling tube or in cracks and crevices of the fermenter walls, must be eliminated by correct design and construction of the fermentation equipment.

Typically, a tank containing 4000 gal of medium may require approximately 4 hr to complete its sterilization cycle when steam at 100 psig is used in the heating-coils. Of this period 90 min is needed to raise the temperature of the medium to 120°C, 30 min at sterilization temperature, and a further 120 min to permit sufficient cooling of the medium before inoculation.

High pressure steam sterilization demands a boiler supplying steam at 100 psig, heating-coils designed to operate at the same pressure and a fermenter constructed to operate at 15 psig. The design must permit air to be thoroughly vented from the equipment in order to eliminate cold pockets that tend to form at the bottom of tall fermenters. The total pressure inside a fermenter is the sum of the partial pressures of the contained steam and air. Any air present will therefore contribute to the total pressure and diminish the partial pressure of the steam, with consequent failure to attain sterilization temperature, as will be appreciated from the figures below.

Absolute pressure of steam (lb/in^2)	Temperature (°F)
14·70	212·00
15·00	213·03
20·00	227·96
25·00	240·07
30·00	250·33

The presence of a small pocket of air, perhaps in a submerged pipe, may have little effect on the temperature of the medium or equipment, but it will cause local superheating or drying of the steam at this point, with consequent failure to sterilize.

When live steam is forced into a cold fermentation medium, heat is transferred and condensate may dilute the medium by as much as 20%. The amount of live steam needed for sterilization may be calculated, assuming the latent heat of steam to be 1000 B.t.u./lb, from the equation

$$W = \frac{10\,V.T}{L},$$

where W is the approximate amount of steam needed (lb); V is the volume of broth to be sterilized (gallons); T is the temperature rise of the broth (°F); and L is the latent heat of steam (B.t.u./lb).

Sterilization in a pressure vessel or autoclave, by contrast, concentrates the medium by about 3–4%. Flasks containing medium and having cotton-wool plugs covered with greaseproof paper are placed in the autoclave, and the door is closed. Injection of steam displaces air from top and bottom of the autoclave until the internal temperature rises to 90–100°C. At this time, the process of venting is stopped by closure of the exhaust valve, but steam continues to enter until 15 psig is registered. The sterilization pressure is maintained until the required sterilization period is completed.

The temperature within a completely vented autoclave is 120°C at 15 psig. Heat is transferred from the steam to the medium in the flasks mainly by external condensation on, and conduction through, the walls of the flasks: little dilution occurs at this stage.

It is normal practice to measure the pressure and temperature of the steam inside the autoclave and to relate the time of sterilization to these two variables. Difficulties sometimes arise when sterilizing large quantities of liquid or liquids containing solid material, and appropriate allowances must be made. This is illustrated in Fig. 5.1. Curve (a) shows the relationship between

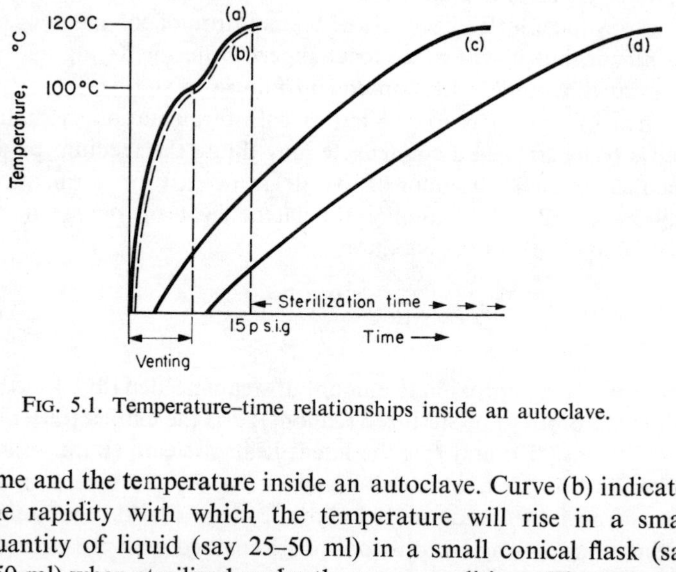

FIG. 5.1. Temperature–time relationships inside an autoclave.

time and the temperature inside an autoclave. Curve (b) indicates the rapidity with which the temperature will rise in a small quantity of liquid (say 25–50 ml) in a small conical flask (say 250 ml) when sterilized under the same conditions. The temperature in a larger quantity of liquid (say 3000 ml) is given by curve (c) and the difference between curves (a) and (c) shows how the temperature of the large volume of liquid lags behind the autoclave temperature: the corresponding temperature–time curve for solids settling out from this larger volume is found in curve (d). It should be clear that, if 15 min at 15 psig is sufficient to sterilize a 50 ml quantity in a 250 ml conical flask, some 2 or 3 hr may be necessary for sterilizing a 3 l. quantity of medium containing solid particles.

During cooling of the autoclave, both the pressure and the temperature of the contained steam slowly decrease. Cooling of the medium occurs largely by evaporation of water, and boiling may occur if the medium loses water too rapidly. Loss of water continues until atmospheric pressure is reached. The door is then opened to permit an inrush of air into the hot autoclave; this aids drying of the outside of the containers, because escape of steam diminishes the partial pressure within the autoclave while its temperature remains at 90–100°C (i.e. the temperature of the residual steam is above its dew point).

The concentration of the ingredients in a sterilized medium will therefore depend upon the technique employed in sterilization, and allowances must be made when changing the scale of industrial operations.

1.2. *Pasteurization and tyndallization*

Simpler but less reliable techniques are sometimes employed for sterilization. Pasteurization is the process of heating liquor to a temperature between 65 and 100°C for a short time, whereby vegetative forms are killed, but most spores are unaffected. The technique is widely used for milk to avoid the deterioration in taste that can occur in the same period of time at higher temperatures. Tyndallization is the process by which vegetative forms are killed at 80–100°C before allowing the resistant spores to germinate at 37°C; the germinated spores are finally killed by reheating to 80–100°C. The cycle is usually repeated. Tyndallization is more reliable than pasteurization, but even this can give no guarantee of sterility. Tyndallization is useful in localities where high pressure steam sterilization is impossible for lack of equipment, or when high pressure steam would bring about decomposition in the medium and render it unusable.

1.3. *Dry heat*

Sometimes it is necessary to sterilize equipment by dry heat, but as early as 1881 Koch showed that this method was relatively inefficient. The efficiency of sterilization improves with increase in moisture content, probably owing to the greater heat stability of proteins in the dry state. Some figures for the coagulation of egg albumen illustrate this point.

Water content (%)	Coagulation temperature (°C)
50	56
18	80–90
6	145
None	160–170

Sterilization by dry heat is used when removal of condensate is difficult (e.g. in the heat sterilization of air). Practical sterility is obtained in an air supply if the temperature is raised to 330°C for 1·15 sec. The relationship between the rate of destruction of bacterial spores and temperature is given by the classical Arrhenius equation:

$$\frac{d \log_e k}{dT} = \frac{E}{RT^2},$$

$$\log_e k = \frac{-E}{RT} + \text{const.},$$

$$\log_{10} k = \frac{-E}{2{\cdot}3\, RT} + A, \tag{5.1}$$

where T is the absolute temperature (°K); E is the energy of activation of the chemical change within the spore (cal/mole); R is the

universal gas constant (cal/mole°C); and k is the velocity constant at temperature T°K of the reaction in which spores are killed (sec^{-1}); A is a constant.
Using spores of *Bacillus globigii*, it has been found that the energy of activation is relatively low (11,000 cal/mole, approx.) and the value of A is 5·26, thus supporting the view that hot air sterilization can be described as a first order chemical reaction.

The death or destruction of micro-organisms is a first order reaction, dependent at any particular temperature only upon the number, N, of organisms present: the process may be represented by the differential equation:

$$\frac{dN_t}{dt} = N_t k.$$

On integration:

$$N_t = N_0 \exp^{-kt},$$

where N. is the number of bacteria present at time t, and N_0 is the number of bacteria present at $t = 0$.

$$\text{Hence } k = \frac{(\log_{10} N_0 - \log_{10} N_t) \times 2{\cdot}303}{t}$$

$$= \frac{(2 - \log_{10} P) \times 2{\cdot}303}{t}, \qquad (5.2)$$

where P is the percentage of organisms surviving the heat treatment after time t: this is numerically equal to 100 N_t/N_0.
Equation (5.1) shows that linear changes in temperature will result in logarithmic changes of reaction rate k, and eq. (5.2) requires an even greater change in the percentage figure P. A small change in temperature with critical range of 270–330°C can therefore lead to a large change in the value of P. Thus when sterilizing air by heat it is always necessary to use accurate temperature recorders and controllers and to maintain the temperature some 5–10°C above the calculated value.

1.4. *Chemicals*

Many chemicals, and especially those able to react with proteins, have been used for disinfection and sterilization. Examples include phenolic compounds, halogens, hydrogen peroxide, ethylene oxide and formaldehyde: all are toxic to man at high concentrations. Chemical sterilants are usually unreliable for sterilization of equipment and except β-propiolactone they are mostly unsuitable for sterilizing complex organic media. Since the time of Lister, 10% phenol or carbolic acid has been used to disinfect wounds and instruments: it forms complexes with and then coagulates protein so that it must be handled with care if phenol burns are to be avoided.

Chlorine water, bleaching powder and other products containing loosely bound halogen are used for disinfecting tanks and pipe-lines in breweries and dairies. The bactericidal action of chlorine depends upon its rapid and intense reaction, even in high dilution, with organic compounds.

Formaldehyde is used for fumigation of rooms and buildings, but it has a pungent smell: its efficiency is impaired by the poor penetration of its vapour and by polymerization to an inert form. It is notable that although *Staphylococcus albus* is sensitive to formalin, several hours' exposure to it are needed to kill cultures that have been dried on cotton threads drawn from a gelatin solution.

Ethylene oxide $\underset{\diagdown O \diagup}{CH_2—CH_2}$ is a useful disinfectant, but its mixtures with air (3–80%) are violently explosive. This substance is toxic at concentrations undetected by smell.

β-Propiolactone (0·5–1·0% v/v) is a toxic material useful for sterilizing whole blood plasma, viral vaccines, and surgical instruments, because it may be rapidly and completely hydrolysed (60°C for 35 min) to the non-toxic β-hydroxypropionic acid.

1.5. *Radiation*

Electromagnetic radiation, especially in the ultraviolet region between 2800–2400 Å, is lethal to micro-organisms that can be penetrated so as to induce irreversible change in the protoplasm. Curtains of ultraviolet light successfully limit dispersal of pathogenic bacteria in hospitals, but spores, especially those of *Bacillus subtilis*, are notoriously resistant to this type of radiation.

High-energy ionizing radiation (γ rays), emitted by the radioactive isotope ^{60}Co is used to sterilize some disposable equipment. The lethal effect is proportional to the radiation dose, but ability of the waves to penetrate the target is inversely proportional to the density of the medium, so that the efficiency of sterilization rapidly diminishes as the path of penetration is increased.

1.6. *Filtration*

Two types of filtration are used for sterilizing liquids and gases, and each employs a different principle. True filtration sterilizes the liquid or gas by passage through a membrane of effective pore size smaller than the bacteria to be removed: these filters include unglazed porcelain candles and sintered glass. The second method does not rely upon diminutive pore size, but rather on the random chance of an organism hitting, and then adhering to, a solid fibre: such filters contain compressed fibrous material (e.g. slag-wool, cotton-wool, or glass-wool) and are mainly used for sterilization of air.

Air can be obtained sterile by passing it at a velocity less than 1 ft/sec through a slag-wool pad packed to a density of 17 lb/ft^3 with fibres of diameter 6μ or less. Almost any fibrous material can provide sterile air when used in this manner, given random packing of the fibre and a sufficient drop in pressure across the filter, but the smaller diameter fibres are always more efficient.

Removal of moisture or oil must be ensured, because all fibres tend to become saturated quickly, and the droplets of moisture would permit infecting organisms to be blown through the filter. There are three common causes of liquid contamination in air filters.

(i) Incorrect design of the compressed-air system.

(ii) Excessive fall of pressure, accompanied by a lowering of the temperature in the air-line supplying the filter.

(iii) Poor design or inadequate maintenance of the air-compresser.

The fundamental problem in design of compressed-air filters arises when air with a 50% relative humidity at 45°F is isothermally compressed to 20 psig, with resultant condensation of water. Two solutions are possible: the compression may be done adiabatically (i.e. without loss of heat), giving a subsequent temperature rise in the compressed air, or the latter may be cooled to remove condensate before being heated again to a temperature exceeding its dew point. A suitable design for a sterile air supply is shown in Fig. 5.2.

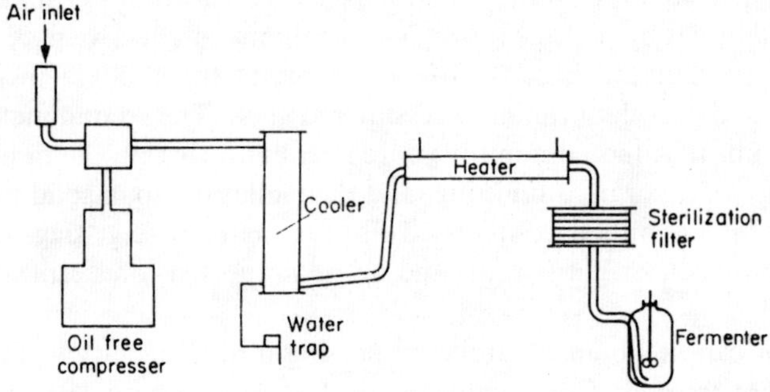

FIG. 5.2. Design of a compressed air system to supply sterile air.

2. Sterility testing

Sterility tests on a small sample of medium may show that the bulk does not seem to be infected, but they cannot prove that it is sterile: such tests have nevertheless practical utility. For example, infection detected in a sterility test on a 500 ml sample of 20,000 l. of "sterilized" bulk medium means that the liquor must have contained at least one living organism. The contamination may, in fact, be much higher: the bulk medium would need to contain at least 40,000 uniformly distributed organisms to ensure detection of the infection in a 500 ml sample; but the infection may be concentrated in a pocket. Similarly, interpretation of sterility tests must be related to the method used, and these in turn must be devised to support growth of contaminants expected in particular fermentations. For example, sterility tests in highly aerobic fermentations must be designed to detect aerobic rather than anaerobic contaminants.

Slow-growing industrial organisms (e.g. actinomycetes and fungi), particularly when infected in the early stages of growth, may be expected to suffer most from the presence of contaminants. The most usual contaminating organisms in actinomycete broths producing streptomycin and in fungal broths producing penicillin are fast-growing spore-forming bacteria, such as *Bacillus subtilis*. This organism may be detected by streaking the broth on the surface of sterile agar medium (beef extract 0·3% w/v; peptone 0·5%; agar 1·5%; pH 6·8) before incubating at 37°C for 3 days, or by inoculating 10 ml of sterile liquid medium (yeast extract 0·5% w/v; peptone 1·0%; Na_2HPO_4 0·75%; glucose 1·0%; Lab Lemco 0·5%; pH 7·0) with 1 ml of the suspect fermentation broth before incubating in the same manner.

3. Asepsis

The principle of asepsis employs a directed change to an unfavourable environment to prevent or retard the growth of contaminating organisms, instead of killing them as in sterilization. Degrees of asepsis occur in all facets of life, and some are particularly important in the food industry: these will now be considered.

3.1. *Cleanliness*

Material suitable for growth of undesirable micro-organisms may sometimes, as with milk products, be excluded by cleanliness.

Milk can support the growth of a variety of contaminating micro-organisms: tanks and pipe-lines for milk products are therefore designed to permit thorough cleansing and are constructed from non-rusting materials with high surface polish and freedom from cracks and crevices, all designed to permit thorough cleansing.

Personal hygiene is another important aspect of cleanliness, particularly in food factories, and in sterile areas where the staff are potential sources of infection. A number of general rules govern personal cleanliness as an aid to asepsis.

(i) Provision of clean clothing from head to toe, preferably including Wellington boots, overalls and a cover for the hair.
(ii) Routine medical examination to ensure freedom from contagious infections.
(iii) Provision of washing facilities and enforcement of regulations for their use.

3.2. *Controlled temperature*

The rate of growth of micro-organisms can be considerably reduced by cooling to 5°C. The growth of most forms is negligible at about 0°C but some of their enzymes probably remain active until a temperature of −80°C is reached. Low temperature alone fails to kill many types of micro-organisms, and spores usually remain viable even after cooling to −80°C; refrigeration (5°C) or deep-freezing (−20°C) may therefore be hazardous if the temperature of the product is permitted to rise temporarily during storage, so that indigenous organisms proliferate but remain undetected because of the subsequent stasis induced by a renewed fall to operational temperature.

3.3. *Controlled humidity*

Water or water-vapour is essential for the continued growth of micro-organisms. Humidity control to prevent condensation is therefore used for industrial storage of fruit: air circulation must be adequate, but not so excessive as to dry the produce, and the temperature must remain uniform. On the other hand, peas may be stored for many months after prolonged drying at normal ambient temperatures, or after drying by the more rapid techniques employing hot air or freeze-drying.

Suggested literature for further reading

1. BOURDILLON, R. B., LIDWELL, O. M. and LOVELOCK, J. E., *Studies in Air Hygiene*, Medical Research Council Special Report, Series No. 262, 1948.
2. MCCULLOCH, E. C., *Disinfection and Sterilization*, Fea & Febiger, U.S.A., 1946.
3. SYKES, G., *Disinfection and Sterilization*, E. & F. N. Spon, London, 1958.
4. WILKINSON, G. R. and BAKER, L. C., Modern trends in steam sterilization, *Prog. ind. Microbiol.* **5**, 237–83 (1964).

CHAPTER 6

Nutrition of Micro-organisms and Media for Industrial Fermentations

1. Nutrition of micro-organisms

The nutritional needs of micro-organisms are diverse, because they differ inherently in ability to synthesize essential growth factors from simple nutrients. However, all micro-organisms demand water, sources of energy, carbon, nitrogen and mineral elements, as well as access to hydrogen and oxygen.

1.1. *Energy sources*

Energy is made available to micro-organisms in several different ways. Some micro-organisms possess photosynthetic pigments that permit utilization of solar energy: these are the photo-autotrophic micro-organisms, including algae and photosynthetic bacteria. Others obtain energy by oxidation of an inorganic substrate (e.g. hydrogen, sulphur and inorganic sulphur compounds, ammonia, nitrites and ferrous salts) that is usually specific for a particular organism: these are the chemo-autotrophic micro-organisms and include bacteria belonging to the genera *Thiobacillus*, *Nitrosomonas*, *Nitrobacter*, *Hydrogenomonas* and *Desulphovibrio*. Finally, we must note the large group of heterotrophic micro-organisms that gain their energy by katabolism of organic substrates and, in particular, of sugars.

1.2. *Sources of carbon*

Carbon is needed for the synthesis of cell material, but the form in which it can be used is closely related to the means employed by the particular micro-organism for supply of energy. Heterotrophic micro-organisms satisfy most of their requirements for carbon by incorporation of metabolites obtained from breakdown of the organic substrates that also provide their energy. Autotrophic micro-organisms, on the other hand, are able to induce the photolytic splitting of water, and this permits fixation of carbon dioxide and its utilization as the sole source of carbon. The two atoms of hydrogen derived from the splitting of each molecule of water provide the enzymic reducing capacity necessary for fixation of carbon dioxide as glyceraldehyde-3-phosphate, and the oxygen atom remaining is either liberated as molecular oxygen (e.g. by algae and higher plants) or is transferred to an oxidizable, inorganic substrate, containing hydrogen or sulphur (e.g. by photosynthetic bacteria). Both these autotrophic processes permit micro-organisms to convert carbon dioxide into glyceraldehyde-3-phosphate: this compound, because of its key role in the carbohydrate metabolism of micro-organisms, is able to provide both energy and the carbon compounds needed for cell growth.

1.3. *Sources of nitrogen*

Micro-organisms show great diversity in their utilization of nitrogen sources. Many micro-organisms are autotrophic for nitrogen, being able to grow on nitrate, ammonia, and sometimes on gaseous nitrogen as sole source of nitrogen, but others need this element to be supplied as amino acids or as purine and pyrimidine bases. Micro-organisms requiring amino acids may be nutritionally deficient, either by their inability to synthesize amino groups or by failure to synthesize certain amino acids: the needs of the former can be met by provision of any one of a wide variety of amino acids as nitrogen source, but the more exacting species demand provision of the specific missing amino acids.

1.4. *Mineral metabolism*

Water is essential for enzymic protein function, for solvation of cellular organic and inorganic materials and as a reactant in many metabolic processes: moreover, in submerged fermentation processes extracellular water provides the suspensory fluid for the cells of the micro-organism, a uniform matrix for distributing nutrients and oxygen and a convenient medium for regulating temperature by rapid heat transfer.

The uptake of water and ions by micro-organisms is regulated by semipermeable membranes, probably composed of lipoprotein films. Small molecules, including those of water, are able to penetrate through the pores of these membranes, but substances of large molecular weight must first undergo digestion by extracellular enzymes. The cell sap has usually a higher ionic concentration than the external medium, so that osmosis determines that water will enter the cells by diffusion through the semipermeable membranes until prevented by turgor pressure (the hydrostatic pressure exerted within the cells by pressure of the expanding protoplast on the semi-rigid cell walls or membranes) or until the external and internal concentrations are equal.

Inorganic ions and other substances of small molecular weight are also able to enter the cells by passive diffusion, but at a slower rate than water. Some micro-organisms are able to speed up this process by operating mechanisms for active transport: these processes are supported by cellular energy, and they permit accumulation of cellular nutrients against a concentration gradient.

Micro-organisms require for their structural or physiological functions, hydrogen, oxygen, phosphorus, sulphur, potassium, calcium, magnesium, sodium and sometimes chlorine, as also the so-called "minor" or "trace" elements, which play important roles as constituents of enzymes or coenzymes; they include manganese, copper, zinc, molybdenum, cobalt and boron, and are usually needed only in relatively small amounts.

1.5. *Accessory food factors*

Many micro-organisms, like animals, are exacting in their nutritional requirements, because they are unable to synthesize certain amino acids, bases and complex organic materials or "vitamins" that are constituents or precursors of enzymes and coenzymes: it is important that these accessory food factors should be supplied in the medium whenever necessary.

2. Raw materials for media

The type of medium used depends upon several interrelated variables. Cost, availability and standardization of the individual ingredients is of prime importance: a cheap raw material is of little use if the world supply is severely limited, and an abundant waste product is equally undesirable if it cannot meet a standard specification.

The second consideration is the percentage composition of the raw material, particularly in available nitrogen and carbon: raw materials containing a large proportion of cellulose have little industrial use, because it is unusual to find organisms showing cellulose activity.

It is also important to relate the medium ingredients to the type of product required. For example, vitamin B_{12} contains cobalt and the fermentation medium must therefore supply this element: similarly, chloride must be included in media used for production of the chlorine-containing antibiotics, chlortetracycline and griseofulvin, and phenylacetic acid or its metabolic equivalent (e.g. *N*-phenylethanolamine) must be provided as a precursor in benzylpenicillin (penicillin G) production.

The physiology of the particular organism employed may determine some of the requisite medium ingredients. *Saccharomyces cerevisiae* strains utilize only hexose sugars, but other yeasts, suitable for food and fodder, are not so limited in their carbon source. Most fungi prefer media containing easily assimilable sources of nitrogen (e.g. cornsteep liquor), but actinomycetes prefer the less readily available sources of nitrogen (e.g. soya bean meal).

Finally, consideration must be given to the balance of the medium constituents, as, for example, in the production of griseofulvin, in which the optimum concentration of nitrogen is related to the nature of the carbohydrate employed.

Production of fermentation products may be considered in two parts—the growth phase of the micro-organism and the phase of product formation by the mature cells. The medium and environment required are seldom the same for these two processes, although the organism itself may sometimes alter conditions during the growth phase to produce an environment suitable for product formation.

The rate at which an organism grows is limited by the number of cells present, by their specific growth rate and by the concentration of any limiting substrate. The rate of product formation is similar to the action of an enzyme upon its substrate:

(i) [Enzyme] + [substrate] $\rightleftharpoons$ [enzyme–substrate complex]

(ii) $\rightleftharpoons$ [Enzyme] + [product_1] + [product_2]

The rate of reaction (i) for any particular concentration of enzyme is given by the Briggs–Haldane or Michaelis equation:

$$v = \frac{V_L S}{S + K_m},$$

where v is the rate of reaction; V_L is the maximum possible rate of reaction with unlimited substrate; S is the molar concentration of substrate; and K_m is the Michaelis constant (equivalent to the substrate concentration permitting half the maximum rate) equal to the dissociation constant of the enzyme–substrate complex (usually of the order of 10^{-3} molar).

The rate of reaction is proportional to the rate of disappearance of substrate (dS/dt), and this, in turn, is proportional to the rate of product formation (dp/dt). In any particular reaction the terminal velocity may be regarded as a constant (k), but the concentration of enzyme will be proportional to the total number of cells (x) producing the enzyme, so that:

$$v = kx \frac{S}{S+K_m}$$

The value of K_m is sufficiently small to be ignored, permitting the approximation

$$-\frac{dS}{dt} = \frac{dp}{dt} = kx.$$

Thus in any particular environment the rate of formation of a product from its substrate is dependent upon the concentration of the cells, and the yield will be limited by the extent to which alternative pathways intervene to produce by-products.

All batch fermentations are subject to a sequence of controlling factors. During the period of growth or development of the organism the medium supplies adequate nitrogen and carbon, but the volume of cells produced is largely determined by the degree of agitation and aeration. During the phase of product formation a variety of different factors may control the yield: limited nitrogen may increase yield in the griseofulvin fermentation; gross deficiency of aeration may be necessary for high yield in the ethanol fermentation; phosphate in excess may decrease yield in the streptomycin fermentation.

Similarly, the pH of fermentation broth has a profound effect upon yield of product. The actinomycetes (especially *Streptomyces griseus*) prefer pH values between 6 and 8 for both growth and antibiotic production, but the higher fungi (e.g. *Penicillium chrysogenum*), although able to grow well at pH 5·0, prefer pH 6·5–7·5 for production of most antibiotics.

Agitation and aeration interact to exert an influence on the optimum concentrations of ingredients. In a nitrogen-limited fermentation the optimum level of nitrogen is sometimes directly related to the degree of agitation or aeration, but the latter is limited in turn by the increase in the thixotropic nature of the broth brought about by increase in cell volume. Fermentations are therefore governed by complex relationships between such factors as nitrogen concentration of the medium, design of the fermenter, degree of agitation or aeration and number of cells produced per unit volume.

The Czapek-sucrose medium, often used for surface cultures of fungi, has the composition: sucrose, 3% w/v; $NaNO_3$, 0·2%; K_2HPO_4, 0·1%; KCl, 0·05%; $MgSO_4$, $7H_2O$, 0·05%; $FeSO_4$, $7H_2O$, 0·001%. Variation in any one ingredient also modifies the properties of the medium in other ways. Alteration of sucrose concentration changes the osmotic pressure, and any convenient method of pH regulation must necessarily alter the concentration of chloride, sulphate, phosphate or other anions. For these reasons, and because it is poorly buffered, this type of synthetic medium has only limited use in industry.

Industrial fermentation media are usually complex. Their compositions are formulated and carefully controlled to a specification often regarded as confidential "know-how": frequently this secrecy implies lack of knowledge about the fundamental biochemistry of the fermentations.

2.1. *Cornsteep liquor*

For a long time cornsteep liquor has been a useful natural additive to synthetic media for improving the growth of micro-organisms or for increasing yields of the desired products.

Cornsteep liquor is a well-balanced source of carbon, nitrogen, sulphur and mineral salts, so that all efforts to attribute its properties to one magic ingredient must fail: nevertheless, the small amount of phenylethylamine in cornsteep liquor, fortuitously acting as a precursor, directed the early penicillin fermentations towards production of benzylpenicillin (penicillin G).

Cornsteep liquor is a by-product when starch and sugar are made from maize. Maize is extracted by steeping in, or by a counter-current flow of, dilute aqueous sulphur dioxide having the twin functions of establishing an acid pH and preventing putrefaction by controlling the indigenous bacterial population. This extraction dissolves substances rich in nitrogen and minerals, and a natural fermentation by thermophilic lactic acid bacteria (*Lactobacillus* spp.) produces lactic acid.

The extracted or steeped corn is drained and filtered before the filtrate is concentrated by heat to about 50% w/v solids concentration; this kills the fermenting bacteria and provides a product that is sold on its solids or nitrogen content.

Cornsteep liquor generally contains about 4% w/v nitrogen and gives, on hydrolysis, a number of amino acids. More than one-quarter of the nitrogen is found as alanine; arginine and glutamic acid account for 8% each; other amino acids include leucine (6%), proline (5%), isoleucine (3·5%), threonine (3·5%), valine (3·5%), phenylalanine (2·0%) and a variety of sulphur-containing amino acids, including methionine (1%) and cystine (1%). Carbon is present as lactic acid and complex polysaccharides resembling gums, in addition to that provided by the amino acids and by small but variable amounts of reducing sugars.

Based upon dry weight, cornsteep liquor contains 1·0% calcium, 2·5% phosphorus and 1·5% potassium. It has also been found to contain riboflavin, niacin, pantothenic acid, biotin and pyridoxine. The precise composition of cornsteep liquor varies with the conditions under which the corn was grown and the process used for extraction.

The acidic nature of cornsteep liquor requires the inclusion in media of up to 1% w/v calcium carbonate to provide a pH suitable for the fermenting organism. A typical medium suitable for production of penicillin G may contain cornsteep liquor to give 0·25% w/v N_2; $CaCO_3$, 1% w/v; KH_2PO_4, 0·8% w/v as a pH buffer; phenylacetic acid, 0·05% w/v as a precursor; lactose, 4% w/v as a slowly utilizable carbon source.

2.2. *Calcium carbonate*

The effect of calcium carbonate on fermentations is not completely explained by its action as an alkali: calcium carbonate used to neutralize the acids in a cornsteep liquor medium gives a high potential for production of penicillin, but the same effect cannot be obtained when potassium or sodium hydroxide is substituted. The particular benefit obtained from calcium carbonate may be due to its reaction with medium ingredients during the sterilization cycle.

Calcium carbonate can exist in two crystalline forms:

(i) calcite in natural limestones or marble;
(ii) aragonite in the shells of molluscs and corals.

Passage of carbon dioxide through cold lime-water precipitates an amorphous form, which soon changes to calcite; if the lime-water is hot, then aragonite is produced instead.

The effect of chalk in fermentation media depends on particle size, the nature of the medium and the sterilization conditions employed. Chalk with a large surface area and small particle size may give an excessively high pH under the same conditions that with larger particles of similar crystallographic composition would give an acceptable pH. Similarly, increased agitation during sterilization accelerates the reactions between chalk and the other medium ingredients, so that a medium within an unagitated 5 l. fermenter giving a post-sterilization pH of 7·0 may develop a pH exceeding 8·0 in the same medium when sterilized in an agitated fermenter.

2.3. *Soya bean*

The soya bean is an annual summer legume growing to more than 6 ft in height and giving a crop of brown seeds borne in pods. The harvested seeds contain up to 40% w/w protein; 18·5%–22% w/w oil; 15% w/w carbohydrate; 5% w/w ash; and water. The ash contains potassium, phosphorus (originally present largely as salts of phytic acid), sulphur, magnesium and iron.

Soya beans are heated to 100°F and flaked before extraction with a solvent that removes the oil, for use in margarine production. The extracted dried material (soya bean meal) can be used as fodder for animals, as a fertilizer, or as an ingredient for fermentation media. At this stage it will contain approximately

8 % w/w nitrogen: this is in much more complex combination and not so readily available to micro-organisms as the nitrogen of cornsteep liquor.

2.4. *Distillers solubles*

The residue after distillation of alcohol from fermented grain or maize contains 6–8 % w/v total solids, rich in protein and the vitamin B complex. The suspended solids are removed by screening and the effluent is concentrated to 35 % w/v solid content to give "evaporator syrup" which is then drum-dried to give "distillers solubles". This, used as an ingredient in fermentation media, supplies nitrogen in the form of protein, together with many accessory food factors.

2.5. *Oils*

Hard fats are chiefly glycerides of palmitic and stearic acids, but the lower melting-point oils are mainly the corresponding glycerides of oleic acid. The three commonest glycerides in fats or oils are the saturated tri-palmitin and tri-stearin and the unsaturated tri-olein.

Fats have higher proportions of carbon and hydrogen than do carbohydrates. Fats on complete oxidation will provide approximately twice the amount of energy and metabolic water that can be derived from an equal amount of carbohydrate.

The vegetable oils can be divided into groups according to the degree of unsaturation.

(a) Oleic or "non-drying" type; these include olive and groundnut oils.
(b) Linoleic or "semi-drying" type: these contain a higher content of the double unsaturated fatty acid, linoleic-acid $C_{18}H_{32}O_2$, found in maize, sunflower and cotton-seed oils.
(c) Linolenic or "drying" types contain linolenic acid $C_{18}H_{30}O_2$,

a fatty acid with three double bonds; they occur in linseed and in soya bean oils.

The "drying" of these oils is a result of oxidation of the unsaturated components on exposure to the atmosphere.

Typical analyses for some oils are listed in Table 6.1. The

TABLE 6.1. COMPOSITION OF SOME TYPICAL COMMERCIAL VEGETABLE OILS

Oil	Saponification value	Iodine value	Percentage saturation	Chief component acids (% w/w)		
				Oleic acid	Linoleic acid	Linolenic acid
Olive	189–195	80–85	9–20	65–84	4–9	–
Groundnut	189–196	85–98	18	56–65	17–26	–
Maize	188–193	117–130	12	45–47	40–42	–
Sunflower seed	186–194	127–136	7–10	30–35	55–65	–
Cotton seed	191–196	103–111	25	25–30	45–50	–
Linseed	189–196	170–185	10–15	15–25	15–20	45–55
Soya bean	190–193	124–133	12–13	25–30	50–55	5–8

saponification value of an oil is the amount (mg per g of oil) of caustic potash needed to saponify the oil: this value is a measure of the ester content of the oil. The iodine value is a measure of the degree of unsaturation as determined by the number of molecules of iodine that can be absorbed by the oil.

Commercial vegetable oils (e.g. maize oil) are usually mixtures of mixed glyceride esters of fatty acids. They may play two different roles in industrial fermentations: they may be used in conjunction with surface-active agents as "anti-foams", or alone as a nutrient source of carbon.

Mineral oils, usually hydrocarbons of high molecular weight, can sometimes be employed in a similar manner as anti-foams but they are not normally metabolized by micro-organisms. Mineral oils therefore tend to accumulate in fermentation broth, and their addition must be regulated to avoid interference with subsequent processes of extraction.

2.6. *Pharmamedia*

This is a clean, yellow, finely ground powder made from the embryo of cotton seed; it contains 56% w/w protein, 24% carbohydrate, 5% oil and 5% ash (containing calcium, iron, chloride, phosphorus and sulphate). It is used commercially in media for production of tetracycline and some of the newer penicillins.

2.7. *Barley*

This is the brewer's traditional raw material for beer manufacture. Barley has a high content of carbon, mainly present in the starch, but the amount of nitrogen is relatively small. Barley of malting quality contains no more than 1·5% w/w nitrogen: a higher nitrogen content would give haze due to deposition of proteinaceous material in the beer.

Barley is allowed to germinate under controlled conditions to permit partial digestion of starch by the enzyme amylase: this process is called "malting". The malted barley is mashed by warming with water and then sterilized, after addition of hops, by boiling in copper pans to give the "wort" or medium to be fermented with a suitable strain of yeast to produce beer.

2.8. *Grapes*

Wine is made by fermentation of grapes usually grown on a light soil anywhere in the warmer temperate zones. The fully ripe grapes are collected, and the "must" or juice is extracted by pressing and fermented either by the yeasts already present on the grapes or by adding a selected strain of wine yeast. The must contains 17% w/w sugar (glucose plus fructose), 1% acid (mainly D-tartaric together with a little malic), and 0·3% ash, containing predominantly potassium and phosphorus. The sugar of grapes is almost completely fermented to alcohol, but some residual pectin is believed to contribute to the smoothness of wine. Again, the best grapes are those low in nitrogen, because excess may result in an undesirable fermentation.

2.9. *Carbohydrate sources*

2.9a. *Lactose.* Lactose or milk sugar is the one useful carbohydrate derived from animals. This sugar is only slowly metabolized so that it does not depress the pH of the medium when present in excess: concentrations of the order of 7–10% w/v can be tolerated in media that would need the glucose or sucrose concentration to be limited to about 0·5% by intermittent additions.

Crude lactose in the form of whey powder (which also contains 4% w/w nitrogen) can be used to replace pure lactose; even so, the price and availability cannot usually make it compete with the carbohydrates produced by photosynthesis. Lactose is therefore a useful carbohydrate source for laboratory fermentations, but the current tendency is to replace it in industrial fermentations by glucose or sucrose fed semi-continuously.

2.9b. *Sucrose*. Sucrose is the common sweet sugar found in sugar cane (*Saccharum officinarum*) and in sugar beet (*Beta alba*). Sugar beet is grown in most temperate zones in Europe, Canada, and the United States, but sugar cane is usually confined to tropical and sub-tropical regions including the West Indies, British Guiana, Cuba, Natal, South Queensland and New South Wales.

Many grades of sucrose are suitable for inclusion in fermentation media.

(i) Pure white sugar, such as "Analar" grade sucrose or ordinary table sugar.
(ii) Brown sugar, normally, but not invariably, cheaper than pure white sugar. Brown sugar is usually the product of the "raw cane" factories, whereas white sugar is produced in the sugar refinery; in the United Kingdom, crude brown sugar is imported to be refined into pure white sugar and a small quantity of edible brown sugar.
(iii) Molasses are the crudest forms of sucrose, being the concentrated mother liquors after crystallization from sugar solutions. Two types of molasses are named according to their vegetable origin: beet molasses and cane molasses.

Beet molasses has approximately the composition 48·5% w/v

sucrose, 1·0% raffinose, 1·0% invert sugar, 10·8% ash, 20·7% organic non-sugars, 18·0% water, 1·5–2·0% nitrogen.

Cane molasses differs in composition from beet molasses mainly in its content of invert sugar; 33·4% w/v sucrose, 21·2% invert sugar, 9·8% ash, 19·6% organic non-sugars, 16·0% water; no raffinose.

Both cane and beet molasses vary in composition from year to year and from locality to locality. It is normal practice for the buyer to select material by experiment during the season and then to buy the year's supply for storage in bulk.

2.9c. *Starch.* Starch is the chief reserve carbohydrate of plants, and it is therefore widely distributed in the vegetable kingdom. The carbohydrate of soya bean meal, groundnut meal, oat flour, rye flour and of maize, wheat and barley is mainly starch. Most commercial starch is made from maize or Indian corn, yielding cornsteep liquor as by-product: subsidiary sources include the tapioca starch obtained from the manihot (cassava) plant and the potato starch derived from the common tuber. Both tapioca and potato starches are relatively difficult to prepare, because of the high water content of the raw plant tissues and susceptibility to spoilage.

Starches are mixtures of two structural polymers: amylose, consisting of straight chains of glucose molecules, and the branched-chain structure known as amylopectin. Glycogen or animal starch is predominantly amylopectin.

Starch is difficult to handle owing to its insolubility in water, but it can be readily hydrolysed by dilute acids, or by enzymes, to give high yields of water-soluble glucose suitable for use in fermentation media.

It is common practice to sterilize glucose separately from the other medium ingredients, because it reacts with ammonium ions and many amino acids to form black, nitrogen-containing compounds: other common sugars, such as sucrose, do not suffer from this limitation.

Dextrins, intermediate in molecular weight between starch and glucose, can be produced by incomplete but controlled acid or enzymic hydrolysis of starch: they are useful as a soluble form of "starch" in small-scale fermentations.

2.10. *Water*

Clean water of dependable composition is essential in large quantities (100,000 gal or more per week for industrial plants) for batching and make-up of fermentation media. Factors that need to be standardized include pH, permanent and temporary hardness, total dissolved solids, colour, clarity and chloride and organic acid contents.

2.11. *Synthetic ingredients including trace elements*

In general, adequate amounts of trace elements are present in media prepared from complex natural ingredients, but they must be included when synthetic media are used.

Many fermentations are supplied continuously with additional nitrogen in the form of ammonia gas, ammonium sulphate, ammonium acetate, ammonium nitrate or urea, and additional carbon may be supplied as glucose, lactate, acetate or alcohol. Care is needed with these additions; all can have a marked effect on pH. Ammonia gas can be added as an alkali to increase pH, and simple carbon sources that are rapidly metabolized may be used to depress pH. Ammonium sulphate may be used to provide nitrogen with little effect on pH if both ammonium and sulphate ions are simultaneously metabolized, but the pH will fall if the ammonium ion is used preferentially.

The acetate ion is quickly metabolized; but nitrate has to be reduced to ammonia: thus ammonium acetate may act as an alkali but ammonium nitrate may behave as an acid. Use of urea can be regarded as a convenient neutral application of ammonia: it is readily metabolized by micro-organisms, but free ammonia is formed during steam sterilization:

$$CO(NH_2)_2 + 2H_2O \rightleftharpoons (NH_4)_2CO_3$$

2.12. *Sources of accessory food factors (vitamins)*

Table 6.2 lists some of the most important accessory food factors that have been shown necessary for certain nutritionally exacting micro-organisms. Column 4 of the table indicates substances that may be needed to rectify metabolic deficiencies, and the commercial sources of raw material for the vitamins are given in column 5.

TABLE 6.2. ACCESSORY FOOD FACTORS SHOWN TO BE NECESSARY FOR CERTAIN NUTRITIONALLY EXACTING MICRO-ORGANISMS

Accessory growth factor	Active form	Chemical group transferred	Substance needed to fulfil metabolic requirement	Raw material source of growth factor
Thiamin (vitamin B_1)	Thiaminepyrophosphate	Decarboxylation and C_2-aldehyde groups	(i) Pyrimidine (ii) Thiazole (iii) Pyrimidine + thiazole (iv) Thiamine	Rice polishings Wheat germ Yeast
Riboflavin (vitamin B_2)	(i) Flavin mononucleotide (ii) Flavin adenine dinucleotide	Hydrogen Hydrogen	(i) Riboflavin	Cereals Cornsteep liquor
Pyridoxal (vitamin B_6)	Pyridoxal phosphate	Amino group and decarboxylation	(i) Pyridoxine (ii) Pyridoxamine or pyridoxal (iii) Pyridoxal phosphate	*Penicillium* spent mycelium Yeast Rice polishings Cereals Wheat seeds Maize seeds Cornsteep liquor
Nicotinic acid or nicotinamide	(i) Nicotinamide adenine dinucleotide (ii) Nicotinamide adenine dinucleotide phosphate (iii) Nicotinamide mononucleotide	Hydrogen Hydrogen Hydrogen	(i) Nicotinic acid or nicotinamide (ii) Nucleotides of nicotinamide	*Penicillium* spent mycelium Wheat seeds Liver
Pantothenic acid	Coenzyme A	Acyl group	(i) Pantothenic acid	Beet molasses *Penicillium* spent mycelium Cornsteep liquor
Cyanocobalamin (vitamin B_{12})		Carboxyl displacement Methyl group synthesis	(i) Cyanocobalamin (ii) Other cobalamins	Activated sewage sludge Liver Cow dung *Streptomyces griseus* mycelium Silage Meat

TABLE 6.2.—*Continued*

Accessory growth factor	Active form	Chemical group transferred	Substance needed to fulfil metabolic requirement	Raw material source of growth factor
Folic acid	Tetrahydrofolic acid	Formyl group	(i) Folic acid (ii) Para-amino benzoic acid	*Penicillium* spent mycelium Spinach Liver
Biotin	Biotin	CO_2 fixation	(i) Biotin	High test cane molasses Cornsteep liquor *Penicillium* spent mycelium
Lipoic acid	Lipoic acid	Hydrogen and acyl groups	(i) *a*-Lipoic acid or thioctic acid	Liver
Purines	Purine nucleotides		(i) Purines (ii) Nucleotides derived from purines	Meat Dried blood
Pyrimidines	Pyrimidine nucleotides		(i) Pyrimidines (ii) Nucleotides derived from pyrimidines	Meat
Inositol	Phosphatides		Inositol	Cornsteep liquor
Choline	Phosphatides		Choline	Egg yolk Hops
Hemins	Cell hemins	Electrons	Hemins	Blood

TABLE 6.3. RAW MATERIALS FOR INDUSTRIAL FERMENTATIONS

(a) *Supplying nitrogen*

Soya bean meal Groundnut meal Pharmamedia	8% N w/w
Cornsteep liquor Whey powder	4·5% N w/w
Oat flour Rye flour Barley Beet molasses Cane molasses	1·5–2·0% N w/w

(b) *Supplying carbohydrate*

Starch	Oat flour Rye flour Barley Soya bean meal Groundnut meal
Sucrose	Pure white sugar Crude brown sugar Beet molasses Cane molasses
Glucose	Pure glucose monohydrate Hydrolysed starch
Lactose	Pure lactose Whey powder

Suggested literature for further reading

1. Symposium on autotrophy:
   I. Introduction, Starkey, R. L., 142–4.
   II. The comparative physiology of autotrophic bacteria, Umbreit, W. W., 145–50.
   III. Recent developments in photosynthesis, Clayton, R. K., 151–64.
   IV. Some thoughts on the energetics of chemosynthesis, Lees, H., 165–7.
   V. Carbon dioxide fixation and substrate oxidation in the chemosynthetic sulfur and hydrogen bacteria, Vishniac, W. and Trudinger, P. A., 168–75.

   Biol. Rev. **26,** 142–75 (1962).
2. *The Bacteria:* a treatise on structure and function.
   III. *Biosynthesis:*
   (i) Assimilation of carbon dioxide by heterotrophic organisms. Wood, H. G. and Stjerernholm, R. L., 41–118.
   (ii) Inorganic nitrogen assimilation and ammonia incorporation, Mortenson, L. E., 119–66.
   (iii) Pathways of amino-acid biosynthesis. Umbarger, E. and Davis, B. D., 167–252.
   (iv) The synthesis of vitamins and coenzymes. Morris, J. G., 253–94.
   (v) Biosynthesis of purine and pyrimidine nucleotides, Magasanik, B., 295–334.

   Academic Press, New York and London, 1962.
3. ATKIN, L., *Yeast Growth Factors, Wallerstein Advances in Beer Quality,*– Part II, Wallerstein Laboratories Communications **25** (87) 113–21, 1962.
4. KERR, R. W., *Chemistry and Industry of Starch*, Academic Press, New York and London, 1944.

5. Kirschenbauer, H. G., *Fats and Oils: an outline of their chemistry and technology*, Reinhold, New York, 1944.
6. Markley, K. S. and Goss, W. H., *Soybean Chemistry and Technology*, Chemical Publishing Co., New York, 1944.
7. Preece, I. A., *Biochemistry of Brewing*, Oliver & Boyd, Edinburgh and London, 1954.

CHAPTER 7

Agitation and Aeration of Microbial Fermentations

MANY designs of fermentation equipment have been used for producing microbial metabolites. The fermenter or vessel in which the process is conducted may be a static test tube for liquid cultures of bacteria, a culture flask containing a mould that grows as a mat on the surface of nutrient fluid, a shake-flask containing micro-organisms growing submerged in a small quantity of fermentation medium agitated by mechanical shaking of the flask, or various shapes and sizes of mechanically agitated and aerated mixing vessels that permit submerged microbial growth in the fermentation medium.

The function of the fermenter is to provide an optimum environment for producing the required metabolic product in maximum yield. This will necessitate providing utilizable nitrogen, carbon, mineral elements and accessory food factors, in a medium aerated in correct proportions at the right time and maintained throughout at optimum values of pH, redox potential and temperature. The metabolic processes occurring in a fermenter are complex, and they are conditioned by the agitation and aeration system employed.

Experiments on a production scale in industrial microbiology are limited and costly: it is therefore usual to construct a model of the production unit so that a large number of experiments can be done without loss of production capacity. It is essential that the small-scale operating conditions of this "pilot plant" (see Chapter 2) should be related to the conditions occurring in the large-scale production plant. Therefore it is necessary to discuss the factors affecting agitation and aeration in vessels of different types and sizes.

Exact similarity in performance between mixing-vessels that differ in size is impossible, and the model-maker must make a number of assumptions before he gets to work. It is usual to ensure that in all fermenters, regardless of size, the ratio of all relevant dimensions is maintained the same: then, despite incomplete knowledge of the mechanisms involved, it is still possible to make progress by applying the mathematical techniques of dimensional analysis. Such application to increased scale of agitation permits calculation of important variables (power input, agitation speed and diameter of the agitator), but this can be done only for fully baffled vessels in the region of turbulent flow. The presence of vortices due to incomplete or non-existent baffling poses problems difficult to solve.

The contribution (in terms of power input) that the flow of air makes to agitation will vary with the size of vessel, becoming more important with increased depth of fermentation broth. Simultaneously, the air bubbling through the fermentation medium will tend to diminish the power contribution from the agitator. The terms K_L (the mass-transfer coefficient for the liquid film) and a (the total interfacial surface area of the gas bubbles) can be used, up to a maximum limit, to express the efficiency of aeration (K_La). Values of K_La can, in turn, be related to both power input to the liquid and to the rate of air flow under any particular conditions. The manufacturer must therefore determine approximate conditions for maximum yield on a small scale, transfer these results to the large scale and then re-examine the relevant variables in the production plant, using knowledge previously gained in the pilot plant.

1. Static cultures

One method of fermentation involving no problems of power

input is the static or surface-culture method used in early work on mould products.

Figure 7.1 shows a typical mould-culture flask suitable for surface cultures on liquid medium of depth up to about 2 cm. Any sterile container providing a suitably shallow layer of medium is

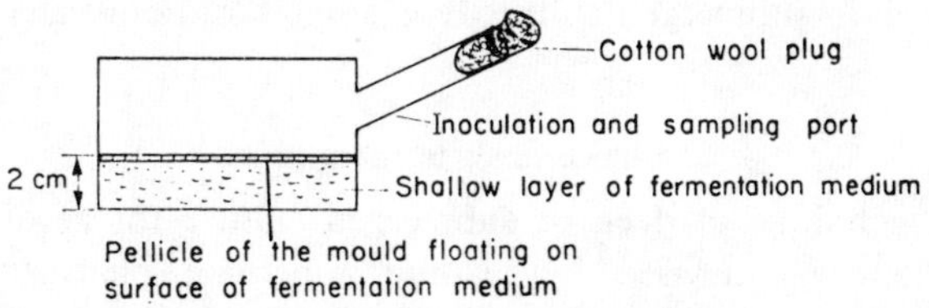

FIG. 7.1. Static mould-culture flask.

satisfactory; in the past, production units were built from various types of bottles or trays. The upper portion of the mycelial pellicle receives a surplus of oxygen with a deficiency of nutrient, but conditions are reversed at the lower surface, where nutrient is abundant but the supply of air is restricted. The mould-culture flask therefore provides an almost infinite number of variable fermentation conditions, ranging from nutrient abundance to nutrient deficiency and from plentiful supply of oxygen to partial anaerobiosis. It may be for these reasons that production of mould metabolites is sometimes more readily achieved in mould-culture flasks, albeit usually in lower yield, than under the more specific but perhaps maladjusted conditions of the more sophisticated fermenters. An alternative explanation may be that the mould-culture flask simulates a multi-stage continuous fermentation (see Chapter 19), initial aerobic stages gradually giving place to the later conditions of deficient aeration.

The culture flasks containing the fermentation medium are sterilized, cooled, and then gently inoculated, preferably with spores, so that the inoculum floats on the surface of the nutrient liquid. The spores germinate to form islands of growth, and these fuse to form a flat pellicle that may become convoluted on the surface of the liquid. Disturbance of the culture flasks must be avoided, to prevent the pellicle from sinking.

The surface growth in mould-culture flasks is similar to growth on agar media, spores being formed at the pellicle–air interface and vegetative mycelium at the pellicle–medium interface. This method is therefore useful for culturing fungi that do not usually grow well in submerged culture, e.g. some basidiomycetes.

2. Shake-flask fermentations

The technique of deep or submerged culture for aerobic fermentation was devised by leading American pharmaceutical companies for commercial production of penicillin in the early years of the Second World War. Previously, many experts had believed that moulds could not be grown immersed in an aerated fluid. The shake-flask is a similar technique, permitting laboratory-scale aerobic fermentations that can be done quickly and with small quantities of nutrient, thereby permitting evaluation of mutant strains and investigation of media. Correlation between shake-flask and stirred aerobic fermenters is effected by comparing corresponding values of $K_L a$.

A shake-flask is a conical flask (usually 250 or 100 ml) that can be moved bodily by mechanical means around the circumference of a circle (usually 2 in. in diameter) with a predetermined frequency (usually 220 rev/min) while maintaining unchanged the relative positions of the flask walls and the platform on which the flask is mounted. This swirling motion agitates the liquid without wetting the cotton-wool plug, and the centrifugal force on the fermentation medium displaces the fluid into a relatively thin layer aerated by diffusion through the liquid surface. Shake-flasks of 250 ml capacity usually contain 25–60 ml of medium, the depth of liquid being adjusted according to the amount of aeration desired, but 100 ml shake-flasks do not normally contain more than 10 ml of medium.

The machine used to move the shake-flasks is termed a "rotary shaker": the best-known design is due to S. Paladino, but simpler

PLATE 7.1. The L.H.E. Orbital Shaker for shake-flask fermentations.

and equally effective equipment is now available as the L.H.E. Orbital Shaker (see Plate 7.1).

The L.H.E. orbital shaker supports the shake-flasks, retained by spring-clips, on a light aluminium platform that floats on four large ball-bearings held in cups. A small electric motor rotates a pulley-wheel situated directly below an eccentric fixed to the floating platform. Each point on the platform therefore describes a regular circular movement whose radius can be determined by the eccentric throw. The shaker is represented diagrammatically in Fig. 7.2.

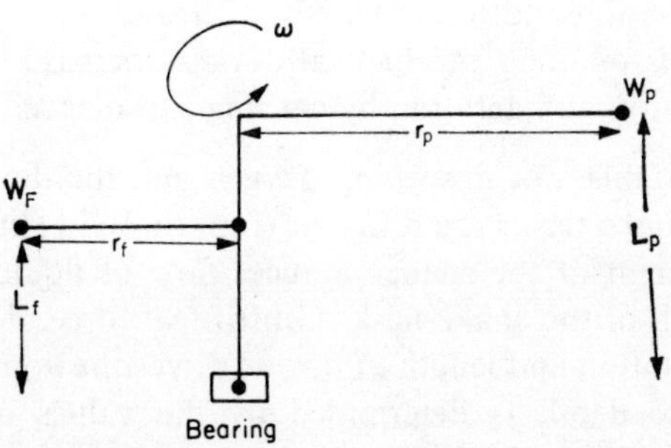

FIG. 7.2. Diagram of the L.H.E. Orbital Shaker, W_p is the effective weight of the platform concentrated at a point distance r_p from the axis of rotation; W_F is the effective weight of the pulley concentrated at a point distance r_f from the axis of rotation. L_p is the distance of plane of rotation of the platform from the bearing; and L_f is the distance of pulley from the main bearing.

The centrifugal force due to motion of the pulley counter-weight (mass of counter-weight × square of the speed of angular rotation × effective radius of rotation) is adjusted to equal that due to motion of the platform, in order to prevent movement of the body of the shaker, so that:

$$\frac{W_p}{g}\omega^2 r_p = \frac{W_F}{g}\omega^2 r_f.$$

The moment about the vertical bearing is given by:

$$\text{centrifugal force} \times (L_p - L_f).$$

The lightest permissible form of construction is attained when the pulley-wheel and the platform are placed close together, making the expression $(L_p - L_f)$ very small.

Agitation and aeration in shake-flasks may be controlled by any of three methods.

(a) Aeration and agitation are inversely related to the depth (and therefore to the volume of fluid) in the shaken flask.
(b) Aeration and agitation are directly related both to the speed of rotation and to the length of throw.
(c) Baffles or other mechanical devices increase the degree of agitation and aeration by causing turbulence.

The mechanism of absorbing oxygen and the flow patterns in unbaffled shake-flasks are different from those in stirred fermenters. Movement of the shaker induces flow of liquid around the interior wall of the shake-flask. Centrifugal force, determined by speed of rotation and length of throw, drives the liquid into a thin layer whose depth is determined by the values of the forces exerted. Air diffuses freely through the cotton-wool plug of the flask to aerate the liquid layer by diffusion, with an efficiency that depends on the depth of fluid.

Other designs of shakers have been tried: a reciprocating shaker with the same travel and the same speed as a Paladino shaker has similar aerating efficiency, but is less useful because medium is splashed onto the wall of the flask so that this eventually becomes covered with a surface culture of the mould. The reciprocating shaker also tends to wet the plugs with medium: if this is permitted, aeration will be reduced, and the sterility of the cotton-wool filter in the neck of the flask can no longer be ensured.

3. Mechanically agitated and aerated fermentations

Figure 7.3 is a diagram representing a simple but typical

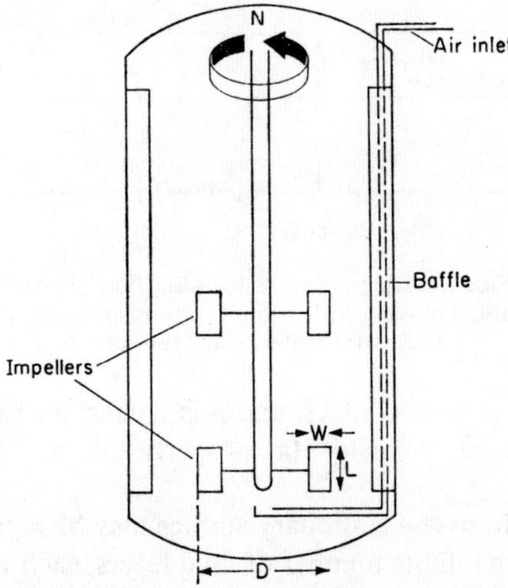

FIG. 7.3. Diagram of a mechanically agitated and aerated fermenter.

mechanically agitated and aerated fermenter that provides a high degree of turbulence. The fermenter is agitated by rotation at N rev/min of two multi-blade turbine impellers of diameter D having blades of width W and height L. Baffles, approximately 1/10th of the diameter of the vessel, are placed vertically on the wall in such a way that at least four are regularly disposed in each fermenter.

3.1. *The theory of agitation*

Agitation is best understood by first considering Newton's Law of Viscous Fluid Flow: this states that the viscous force F opposing motion at the interface between two liquid layers flowing with a velocity gradient dv/dx (see Fig. 7.4) is given by the expression:

$$F = \mu A \frac{dv}{dx}, \tag{7.1}$$

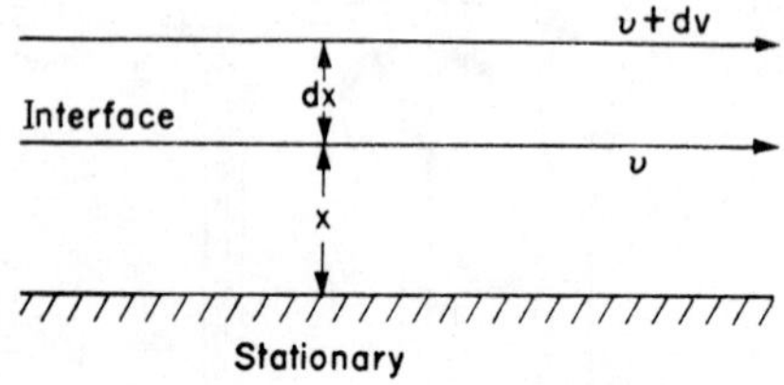

FIG. 7.4. Diagram of "streamline flow" in a liquid moving with a small velocity gradient dv/dx over a stationary surface.

where A is the area of contact, and μ is a constant referred to as the coefficient of viscosity (a property of the fluid under consideration).

Flow of liquid over a stationary surface may be regarded as the movement of an infinite number of fluid layers, each moving with a velocity that increases with distance from the surface, as shown in Fig. 7.4. If the velocity gradient is small, then the layers of liquid will flow over each other without eddy formation: this is "streamline flow". When the velocity gradient is large, local turbulence or eddy formation occurs, and Newton's Law is no longer applicable: this is "turbulent flow" and is illustrated in Fig. 7.5. There is a "critical velocity gradient" between these two extreme conditions, in which streamline flow verges into turbulent flow.

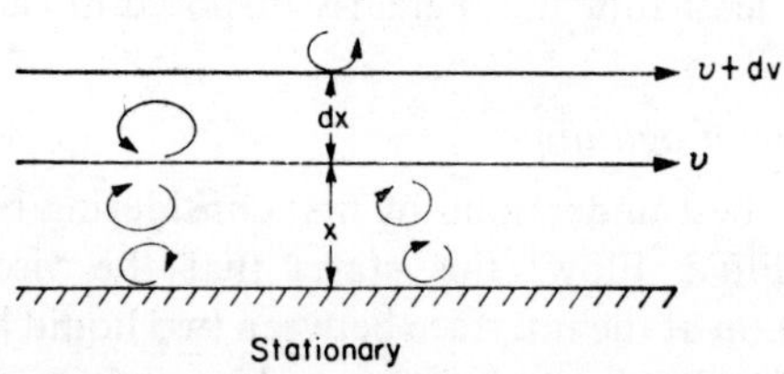

FIG. 7.5. Diagram of "turbulent flow" in a liquid moving with a large velocity gradient dv/dx over a stationary surface.

In the special case of fluid flowing through a pipe it is usual to measure only the mean velocity of the fluid. The particular velocity at which streamline flow changes to turbulent flow is termed the "critical velocity" (V_c) and is a function of the properties of the fluid (the viscosity μ and the density ρ) and of the diameter (d) of the pipe:

$$V_c = f(\mu\rho d). \tag{7.2}$$

It is axiomatic that both sides of an equation for a physical phenomenon must be capable of being represented in the same units (i.e. the equation must be dimensionally homogeneous). For example, in eq. (7.1) we know by definition:

force = mass × acceleration.

Therefore force has the dimensions:

$$\frac{\text{mass} \times \text{length}}{(\text{time})^2}$$

and this may be abbreviated to read MLT^{-2}. It then follows that the right-hand side of eq. (7.1) must have the same dimensions. Substituting the dimensional terms L^2 and LT^{-1}/L for A and dv/dx respectively, it can be concluded that μ must have the dimensions:

$$\frac{\text{mass}}{\text{length} \times \text{time}} \text{ or } ML^{-1}T^{-1}.$$

Similarly, for eq. (7.2) to be dimensionally homogeneous, the right-hand side must have the dimensions of velocity (i.e. LT^{-1}): this can be satisfied when:

$$V_c = N_{(Re)} \cdot \frac{\mu}{\rho d}, \tag{7.3}$$

where $N_{(Re)}$ is the constant of proportionality associated with viscous flow and termed Reynolds number.

Reynolds number is a dimensionless number, specifically

associated with viscous flow of fluids; it can be obtained by rearrangement of the more usual form of eq. (7.3):

$$\text{Reynolds number} = \frac{V\rho d}{\mu}.$$

Expressing this relationship in words we find that:

$$\text{Reynolds number} = \frac{\text{linear velocity} \times \text{density} \times \text{relevant linear dimension}}{\text{viscosity}}.$$

The linear velocity of fluid at the periphery of the impeller in a mixing vessel is proportional to the product of the speed of agitation N and the diameter of the agitator; density and viscosity are self-evident; and the relevant linear dimension is the diameter of the impeller D. Therefore

$$N_{(Re)} = \frac{N\rho D^2}{\mu}. \tag{7.4}$$

When viscous forces are dominant (e.g. in fully baffled vessels), Reynolds number is the valid parameter, but experience shows that when gravitational force is an important factor (e.g. in systems in which a vortex is present), another dimensionless number termed the Froude number $N_{(Fr)}$ must be included:

$$N_{(Fr)} = \frac{DN^2}{g}. \tag{7.5}$$

The power P absorbed by the fluid in a mixing vessel (equivalent to the power P absorbed by the impeller) can be related to the speed of agitation N and the impeller diameter D by the use of another dimensionless group termed the power number $N_{(p)}$:

$$N_{(p)} = \frac{Pg}{\rho N^3 D^5}, \tag{7.6}$$

where g = acceleration due to gravity.

The three dimensionless numbers $N_{(p)}$, $N_{(Re)}$, and $N_{(Fr)}$ can be related, for geometrically similar mixing-vessels, by the equation:

$$N_{(p)} = K\{N_{(Re)}\}^m\{N_{(Fr)}\}^n, \tag{7.7}$$

where K is a constant and m and n are indices of power.

In fully baffled vessels the Froude number can be ignored, hence:

$$N_{(p)} = K\{N_{(Re)}\}^m. \tag{7.8}$$

Therefore
$$\frac{Pg}{\rho N^3 D^5} = K\left\{\frac{N\rho D^2}{\mu}\right\}^m. \tag{7.8}$$

Experimentally, it has been found for fully baffled vessels that, if the log of the power number $N_{(p)}$ is plotted against the log of the Reynolds number $N_{(Re)}$, the picture shown in Fig. 7.6 is

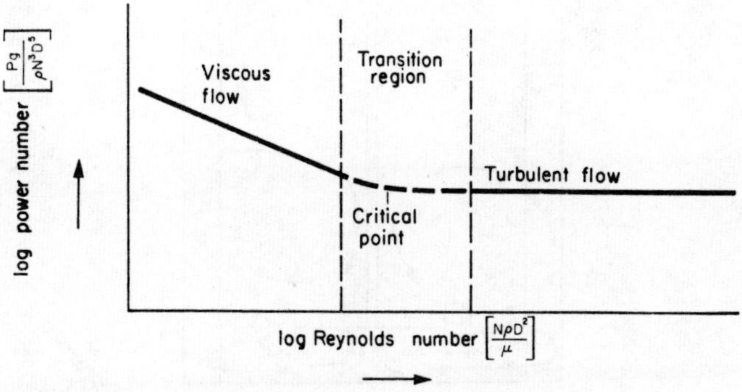

FIG. 7.6. The relationship between the Reynolds number and the power number.

obtained. In the region of viscous flow m has the value -1, so that the power absorbed (calculated from eq. (7.8)) is given by

$$P = \frac{K_1}{g}\mu N^2 D^3. \tag{7.9}$$

On the other hand, the value of m is zero in the region of turbulent flow, so that
$$P = \frac{K_2}{g}\rho N^3 D^5. \tag{7.10}$$

It will be noted that the power absorbed is a function of viscosity in the region of viscous flow, but is a function of density in the region of turbulent flow.

Comparison of eq. (7.10) with eq. (7.9) shows that increased speed of agitation results in proportionately greater increase in power input to a fluid in turbulent flow than to a corresponding fluid in viscous flow. This extra power is absorbed in the eddy currents resulting from the turbulent flow of fluid in a fully baffled system: these are shown in Fig. 7.7. For comparison, the flow pattern in an unbaffled vessel is illustrated in Fig. 7.8.

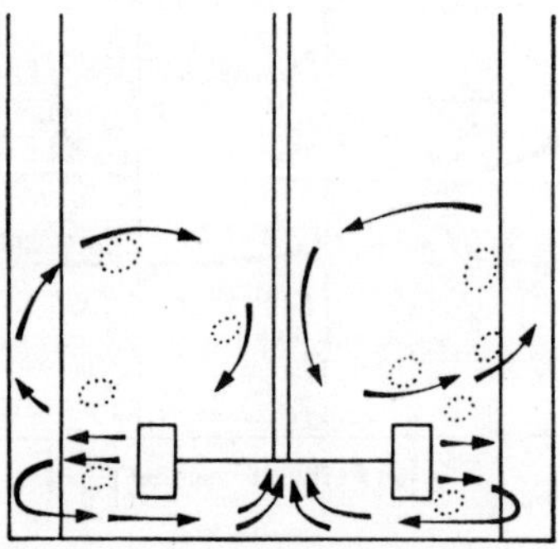

FIG. 7.7. Fluid flow pattern for turbine impellers in a mixer fitted with baffles.

Rotation of the impeller in an unbaffled vessel is accompanied by formation of a vortex in the fluid, and this makes the power absorbed less than that absorbed in a fully baffled system. Nevertheless, power absorbed in an unbaffled vessel continues to increase with speed of rotation until air is sucked into the fluid. Thereafter, any further increase in speed of rotation will diminish the power absorbed, because the impeller now rotates partly in air (see Fig. 7.8): the centrifugal force developed tends to separate heavy and light concentric layers within the fluid.

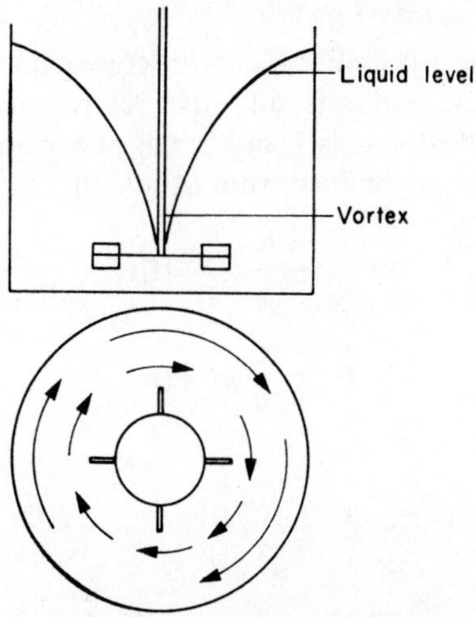

FIG. 7.8. Fluid flow pattern for turbine impellers in an unbaffled mixing vessel.

Vortex production in an unbaffled vessel determines that the power number is related both to the Reynolds number and to the Froude number as given in eq. (7.7):

$$N_{(p)} = K\{N_{(Re)}\}^m\{N_{(Fr)}\}^n.$$

Experimentally, the situation is found to be further complicated, because the plot of log power number against log Reynolds number varies with the agitator speed in the region of turbulent flow and because the Froude number itself is a function of the Reynolds number.

It should now be clear that unbaffled vessels are fundamentally poor mixing-vessels because of the intense vortexing that occurs under certain conditions, and this makes it practically impossible to arrange any rational scale-up from one size to another. On the other hand, performance of baffled vessels in the region of turbulent flow (i.e. with high Reynolds numbers) can be predicted according to size.

3.2. *Scale-up of agitated vessels*

For geometrically similar (i.e. with corresponding dimensions having the same ratio as all other corresponding boundary dimensions) baffled vessels 1 and 2, the power absorbed by fluid in turbulent flow is obtained from eq. (7.10):

$$P_1 = \frac{K_2}{g}\rho N_1^3 D_1^5,$$

$$P_2 = \frac{K_2}{g}\rho N_2^3 D_2^5.$$

Therefore

$$\frac{P_1}{P_2} = \frac{N_1^3 D_1^5}{N_2^3 D_2^5}. \qquad (7.11)$$

It may be assumed that power absorbed per unit volume $(P)/(\text{Vol})$ is a measure of the intensity of agitation: when this value is the same for both vessels

$$\frac{P_1}{P_2} = \frac{\text{Vol}_1}{\text{Vol}_2} = \frac{D_1^3}{D_2^3}. \qquad (7.12)$$

Equating values of P_1/P_2 obtained from eq. (7.11) and (7.12),

$$\frac{D_1^3}{D_2^3} = \frac{N_1^3 D_1^5}{N_2^3 D_2^5}.$$

Therefore
$$\left(\frac{N_2}{N_1}\right)^3 = \frac{D_1^2}{D_2^2}.$$

Therefore
$$\frac{N_2}{N_1} = \left(\frac{D_1}{D^2}\right)^{2/3}. \qquad (7.13)$$

Equation (7.13) shows that, in any two geometrically similar vessels of different size, the power input per unit volume will be the same in both vessels when the inverse ratio of their agitator speeds is the same as the square of the cube root of the ratio of their impeller diameters.

The procedure for scale-up is more complicated when the mixing vessels are geometrically dissimilar. For example, empirical experiments in fully baffled vessels exhibiting turbulent flow have shown that the power number is given by the expression:

$$N_{(p)} = \frac{P_g}{\rho N^3 D^5} = \frac{K(L)}{D}\frac{(W)^{1\cdot 5}}{(D)}\frac{(B)^{0\cdot 8}}{(6)}\frac{(R)^{0\cdot 4}}{(4)}\frac{(J)^{0\cdot 3}}{(D)}, \qquad (7.14)$$

where B is the number of blades on the impeller, R is the number of baffles; J is the baffle width; W is the width of the impeller blades; L is the height of the impeller blades; and D is the overall diameter of the impeller.

Clearly, calculation of proportionate dimensions and agitator speeds to give equal intensities of agitation will be more difficult when vessels are geometrically dissimilar.

3.3. *Power absorbed from aeration*

The power P taken from an impeller is given by eq. (7.10) but this value will diminish as the rate of aeration increases from zero, because the overall density of fluid containing occluded air is reduced. The diminution in power absorbed from the impeller is counteracted to a certain extent by the power supplied by aeration. Therefore, we may say that the intensity of agitation in an aerated fermenter is related to the sum of the power supplied by the agitator and that supplied by aeration.

The work done by aeration may be calculated from Bernouilli's theorem. This states that in any continuous fluid the total energy of every particle is the same. Thus the sum of energies [internal (U), pressure (pV), potential (H) and kinetic $(u^2)/(2g)$] of a particle of air at the inlet to the fermenter must equal the sum of the corresponding energies of a similar particle of air at the exit (more accurately, at the surface of the broth) plus the work done during passage of the air particle through the broth:

$$U_1 + p_1 V_1 + H_1 + \frac{u_1^2}{2g} = U_2 + p_2 V_2 + H_2 + \frac{u_2^2}{2g} + W. \qquad (7.15)$$

The kinetic energy of air at the broth surface in the fermenter is small, so that $u_2^2/2g$ can be ignored. The difference in potential energy $H_1 - H_2$ between particles at the top and bottom of the fermenter is numerically insignificant. The process is isothermal (i.e. at constant temperature), and therefore there is no change in internal energy ($U_1 = U_2$). Work done by an ideal gas in isothermal expansion from p_1V_1 to p_2V_2 is given by the integral of pdv between the limits 1 and 2.

After integration of pdv and addition of the kinetic energy term, the total work done by passage of air through the broth is given by:

$$W = \frac{u_1^2}{2_g} + p_1V_1 \log_e \frac{p_1}{p_2}. \tag{7.16}$$

Equation (7.16) shows that the work done by aeration can be increased both by a decrease in diameter of the air inlet (with a consequent increase in the kinetic energy term $u_1^2/2g$) and by an increase in depth of the fermentation broth (giving an increase in the value of the expression $p_1V_1 \log_e(P_1/P_2)$). These modifications are important when fermentations are agitated solely by aeration (e.g. when using porous candles) or by evolution of metabolic gas (e.g. carbon dioxide from acetone–butanol and from ethanol fermentation broths).

3.4. *Interaction of mechanical agitation with aeration*

Theoretical discussion about the effect of agitation upon aeration can be only approximate, because the physical properties (e.g. viscosity and density) of a fermentation broth alter during the fermentation period. Moreover, the apparent viscosity of many fermentation broths will vary with the type and with the intensity of agitation (i.e. they are non-Newtonian fluids).

Consideration of eq. (7.10) shows that the power absorbed by the broth from the mechanical agitation system depends on the cube of the revolutions per minute and on the fifth power of the agitator diameter: therefore small increments in either or both of these variables will result in a relatively large increase in power input. On the other hand, eq. (7.16) shows that there is no correspondingly large increase in work done per unit time (power input) resulting from increased air-flow, because increase in power input depends mainly on the square of the air velocity (u^2).

It is observed in industrial practice that agitation largely determines the yield of product in an aerobic fermentation. At any one rate of air-flow exceeding a critical minimum, the yield is greatly diminished when agitation is decreased; in contrast, a corresponding diminution of air-flow at constant agitator speed has relatively little effect on yield. For this reason, it has been suggested that the agitation and aeration can be expressed in terms of the *total* power input to each unit volume of broth. This implies that the work done per unit time (power input) on each unit volume of broth by the air-flow must be greater in tall than in short fermenters (see eq. (7.16)) and that the former will require correspondingly less power from the agitator.

The mechanism of the process whereby oxygen passes from the bubbles of air within the broth to the organism may be discussed conveniently in terms of the theory of diffusion through thin films. It is necessary to postulate that a thin film of relatively stationary material exists at the interface between two dissimilar fluids, even though the main bulk may be in turbulent motion. The transfer of material from one phase to another then occurs by two mechanisms—the bulk movement of fluid and the diffusion of molecules through relatively stagnant film.

The oxygen molecules in a bubble of air must traverse several thin films to become accessible to the micro-organisms growing in the broth:

(1) the gas film at the periphery of the air bubble;
(2) the liquid film at the gas–liquid interface;
(3) the liquid film at the liquid–organism interface.

The liquid film at the liquid–organism interface will have a negligible effect on the passage of oxygen, because it has been found experimentally that the respiration rate of most micro-organisms is independent of the dissolved oxygen concentration. There is, however, a critical minimum concentration, and this may be even lower than one-tenth of the saturation concentration for oxygen in broth. Similarly, the effect of a gas film at the periphery of a bubble may be neglected, because it has little effect on the rate of solution of a relatively insoluble gas such as oxygen or nitrogen.

The rate of solution of oxygen into a fermentation broth (N_A) is therefore given by the classical mass transfer equation:

$$N_A = K_L a\,(C_E - C_L) \qquad (7.17)$$

where K_L is the "overall mass transfer coefficient": this can be regarded as a measure of the thickness of the hypothetical stationary liquid film; a is the total area of the gas–liquid interfaces; C_E is the concentration of dissolved oxygen in equilibrium with the partial pressure of the oxygen in the bubble; and C_L is the actual concentration of dissolved oxygen; $C_E - C_L$ is therefore the concentration gradient or driving force.

C_E is, for all practical purposes, equivalent to the saturation concentration of dissolved oxygen, because the partial pressure in the gas bubbles within the fermentation broth does not alter appreciably. The term $K_L a$ may be defined as the "aeration efficiency" of a fermenter, and its value may be determined by the method outlined below.

The concentration of oxygen in a liquid is reduced to a low level when sodium sulphite ($>0{\cdot}05$ M) is oxidized in the presence of a catalyst, such as the ions of copper or manganese. This is the basis of the copper sulphite method for comparing aeration efficiencies in fermenters.

If C_E is the saturation concentration of dissolved oxygen in the sulphate solution resulting from oxidation of the sulphite, then eq. (7.18) gives the rate of oxygen absorption (dc/dt) by the liquid in the vessel under test

$$\frac{dc}{dt} = K_L a(C_E - C_L). \tag{7.18}$$

But we know already that C_L may be neglected; therefore

$$\frac{dc}{dt} = K_L a(C_E). \tag{7.19}$$

The aeration efficiency of a fermenter can also be determined by the gassing-out method. A fermenter filled either with broth or with a salt solution is de-oxygenated by passage of nitrogen. The rate of solution of oxygen is then determined by polarograph when the fermenter is operated under the desired conditions. Integration of eq. (7.18) gives the expression

$$\log_e (C_E - C_L) = -K_L a T + \text{constant}. \tag{7.20}$$

The natural log of $(C_E - C_L)$ is plotted against time (T), and the value of $K_L a$ is determined from the slope of the graph.

Experiments have shown that the mass transfer coefficient K_L for a liquid film is independent of the size of bubble and of the velocity with which the bubble rises: its value can be predicted from the physical properties of the fluid alone. The power dissipated by agitation, on the other hand, has been shown experimentally to effect an increase in the gas–liquid interfacial area, thereby increasing the rate of solution of oxygen. The agitator thus has three functions.

(1) To diminish the size of the bubbles of air and thereby increase the interfacial area available for oxygen transfer.
(2) To maintain the liquid in motion so that dissolved oxygen and nutrients are made available to the micro-organism.
(3) To maintain uniform temperature.

The values of $K_L a$ have been correlated with mechanical agitation power per unit volume for various air velocities at the inlet. Dimensionally similar fermenters will give similar fermentation patterns when operated under conditions of similar aeration efficiency and of total power input per unit volume.

3.5. *Most economic design of a fermenter*

In general, microbial fermentations can be divided into two groups: those requiring high and those requiring low aeration. The micro-aerophilic fermentations (e.g. production of ethanol, 2,3-butanediol, acetone) need only sufficient agitation for temperature control and for adequate exchange of nutrients between the broth and the organism. The energy supplied by compressed air is usually adequate for this purpose, and mechanical stirring is unnecessary except during the sterilization cycle (see Chapter 5).

Fermentations requiring high aeration (e.g. production of sterols and most antibiotics) must be agitated with sufficient power to maintain uniform temperature control, to break up the bubbles of air, to maintain sufficient exchange of nutrients between the broth and the organism and to prevent formation of stagnant pockets of broth (e.g. near the fermenter wall). Moreover, the impellers must be so spaced vertically that their flow-patterns do not interfere, but they must be sufficiently close to avoid formation of stagnant regions in the broth.

During the growth phase of most industrial fermentation processes there is an increase in apparent viscosity (rigidity) of the broth, tending towards the formation of stagnant zones. This will result in decreased turbulence and further diminish the aeration efficiency (K_La).

Experiment has shown that, in conditions of high turbulence, the power P consumed by multi-bladed turbine impellers is related to the dimensions of the impeller and its blades (see Fig. 7.3) and to the density ρ of the fluid:

$$\frac{Pg}{N^3D^5\rho} = \frac{(\text{constant})\ WL\,(D-W)}{D^3}. \qquad (7.21)$$

This means that, at constant agitator speed (N) and with a constant overall diameter of impeller (D), the power absorbed by the impeller is proportional to the volume that it sweeps:

$$P = (\text{constant})\ WL\,(D-W). \qquad (7.22)$$

When the dimensions of the impeller blades are altered to vary the volume swept, it is found that the power absorbed is proportional to the aeration efficiency up to a limiting value. Any increase in power absorbed beyond this value would be wasted.

Unfortunately, the relationship between the diameters of impeller and fermenter giving greatest efficiency of aeration is not known; in industry, impellers are usually about $\frac{1}{2}$ to $\frac{3}{4}$ of the diameter of the fermenter. It is to be expected that the diameter of the impellers should be somewhat increased for broths of high viscosity that tend to diminish turbulence and aeration efficiency.

The optimum distance between impellers is also undetermined, but this approximates in practice to about 1 or $1\frac{1}{2}$ impeller diameters. Again, it might be expected that the distance between the impellers should be diminished as the viscosity of the broth increases.

It may therefore be concluded that, for any given fermentation process, maximum yield will require that the sum of the power per unit volume absorbed by the broth from the agitator (P) and from the air (W) must exceed a certain minimum value; this approximate value is first determined by small-scale experimentation. The contributions to the power absorbed per unit volume made by the air-flow and by the agitator must each exceed minimum values in order that the aeration efficiency ($K_L a$) be raised above its limiting value.

Suggested literature for further reading

1. Bates, R. L., Fondy, P. L. and Carpstein, R. R., An examination of some geometric parameters of impeller power, *I and EC Process Design and Development* **2,** 310–14 (1963).
2. Calderbank, P. H., Physical rate process in industrial fermentation. Part I, The interfacial area in gas–liquid contacting with mechanical agitation, *Trans. Instn. chem. Engrs.* **36,** 443–63 (1958); Part II, Mass transfer coefficients in gas–liquid contacting with and without mechanical agitation, *Trans. Instn. chem. Engrs.* **37,** 173–85 (1959).
3. Calderbank, P. H. and Moo-Young, M. B., The power characteristics of agitators for the mixing of Newtonian and non-Newtonian fluids, *Trans. Instn. chem. Engrs.* **39,** 337–48 (1961).

4. COOPER, C. M., FERNSTROM, G. A. and MILLER, S. A., Performance of agitated gas–liquid contactors, *Ind. Engng. chem.* ind. (int.) edn., **36,** 504–9 (1944).
5. HAMER, G. and BLAKEBROUGH, N., Turbine impellers as gas–liquid contacting devices, *J. appl. Chem.* **13,** 517–24 (1963).
6. LANGHAAR, H. L., *Dimensional Analysis and Theory Models*, Chapman & Hall, London, 1951.
7. MCLAREN-WHITE, A. and BRENNER, E., Studies in agitation, *A.I.Ch.E.Jl.* **30,** 585–96 (1934).
8. RICHARD, J. W., Power input to fermenters and similar vessels, *Br. Chem. Engng.* **8,** 158–63 (1963).
9. RUSHTON, J. H., COSTICH, E. W. and EVERETT, H. J., Power characteristics of mixing impellers, *Chem. Engng. Prog.* **46,** 395–404 (1950).
10. STEEL, R. (ed.), *Biochemical Engineering*, Heywood, London, 1958.

CHAPTER 8

Food Microbiology

MICRO-ORGANISMS play four distinct roles in food microbiology: their cellular material may be utilized as food supplements, their enzymes may be employed as accessory agents in food production, they may be the agents of food spoilage, and some human pathogenic forms may be the cause of "food poisoning". These aspects of food microbiology will now be considered, but consideration of some other topics related to food microbiology will be deferred for subsequent treatment. Thus, production of lactic acid and of alcoholic beverages by fermentation will be described in Chapter 10, vinegar production and the synthesis of L-ascorbic acid in Chapter 17, and fermentations yielding glycerol and glutamic acid in Chapters 9 and 12, respectively.

There remain a number of other processes employed in the food industry that depend upon microbial fermentations for preservation of the product or for the characteristic quality of the food, be it aroma, flavour, appearance or texture, but too empirical for present consideration: these include methods for production of cocoa, cheese, citron, coffee, olives, pickles, *Sauerkraut*, tea and vanilla.

1. Micro-organisms in the production of food

Entire organisms frequently constitute the end-product of microbiological processes in the food and feeding stuff industries. Procedures are relatively simple and depend upon maximum production in the shortest time of microbial tissues having the requisite chemical or biochemical composition. Essentially, the efficiency of these processes is determined by the characteristics of growth of micro-organisms.

1.1. *The growth curve*

Growth or increase in total cell matter is fundamental to biological systems, but its assessment in micro-organisms presents some practical difficulties. The best estimate of growth is probably obtained from values for total cellular nitrogen, but this is inconvenient to determine: for most purposes, determination of the increase in dry weight of the cells is adequate, but it is important to note that this value can be inflated when storage products are being formed. Other useful measures of growth include turbidity of cultures of unicellular organisms (the optical density is a function of the product of cell size and number and can be related by means of a calibration curve to the dry weight) and centrifugation, to determine the volume of packed cells or mycelium. Indirect methods based on determination of cellular activity (cell counts, carbon dioxide or acid production) are sometimes useful, but can be misleading if used indiscriminately.

The logarithm of the dry weight of a unicellular organism growing in submerged culture when plotted against time gives a curve of characteristic shape, the detailed form of which is determined by the type of organism and the environmental conditions. This is the growth curve, and it consists typically of five portions: the lag phase, the acceleration phase, the exponential phase, the retardation phase and the stationary phase (Fig. 8.1).

1.1a. *The lag phase*. For any given micro-organism, the period of time that elapses between inoculation of fresh medium and detection of visible growth is known as the lag phase. It is the period of adaptation to the new environmental conditions: its duration depends on the nature of the micro-organism, the size and physiological status of the inoculum, the amount of medium transferred with the inoculum, the changed concentrations of

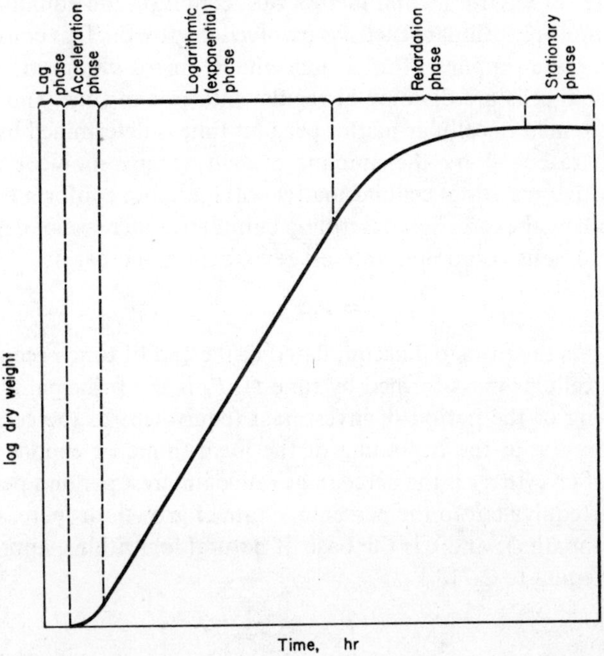

FIG. 8.1. Typical growth curve of a micro-organism in submerged culture.

essential metabolites in the new medium, the temperature and, sometimes, the quality and intensity of light. In general, the lag phase is shortened by increased size of inoculum, by increased temperature up to the optimum and by transfer to a nutritionally more complete medium: it may be greatly diminished or even eliminated by inoculation with a culture that is already in the exponential phase of growth.

1.1b. *The logarithmic or exponential phase*. The lag phase is terminated by a short period of accelerating growth rate before the logarithmic or exponential phase begins, when the growth rate is constant at its maximum for the particular organism and the given set of environmental factors that condition the equilibrium of the multiple cellular reactions involved in growth. This constant growth rate implies, for a unicellular micro-organism, that individual cells are dividing at regular intervals of time, and that the increment in cellular matter per unit time is determined by the growth rate and by the amount present at any moment. The cumulative growth of cellular matter will therefore conform to the mathematical expression describing cumulative increase of capital by continuous compound interest (exponential increase):

$$P = P_0 \mathrm{e}^{rt},$$

where P is the principal accumulated at the end of time t (equivalent to cellular mass formed by time t); P_0 is the principal at the beginning of the period of investment (equivalent to the cellular mass present at the beginning of the logarithmic or exponential phase of growth); r is the percentage rate of interest per unit period of time (equivalent to the percentage rate of growth or increase in cellular matter); and e is the base of natural logarithms, approximately equal to 2·718.

Therefore
$$\frac{P}{P_0} = \mathrm{e}^{rt}.$$

By taking logarithms of both sides of the equation:

$$\log_e \frac{P}{P_0} = rt,$$

therefore
$$\log_e P - \log_e P_0 = rt,$$

therefore
$$r = \frac{\log_e P - \log_e P_0}{t}. \tag{8.1}$$

Equation (8.1) states that the rate of growth (r) is given by the slope of the graph of $\log_e P$ against time (t), so that in general terms

$$r = \frac{\log_e P_2 - \log_e P_1}{t_2 - t_1}. \tag{8.2}$$

It is usually more convenient to plot the growth curve in terms of common logarithms to the base 10: its slope (r_c) is given by eq. (8.3):

$$r_c = \frac{\log_{10} P_2 - \log_{10} P_1}{t_2 - t_1}. \tag{8.3}$$

Since common logarithms may be converted to Napierian logarithms by multiplying by 2·3026, we may substitute $2{\cdot}3026r$ for r_c in eq. (8.3) to obtain the growth rate r in terms of common logarithms:

$$r = \frac{\log_{10} P_2 - \log_{10} P_1}{2{\cdot}3026(t_2 - t_1)}. \tag{8.4}$$

Sometimes it is convenient, especially with unicellular micro-organisms, to use logarithms to the base 2 for the growth curve plot, because one division on the ordinate scale then represents a doubling of the population or an increment of one generation. The slope of this graph (r_u) is equal to the reciprocal of the generation time (g), the time taken for each cell division, so that

$$r_u = \frac{1}{g} = \frac{\log_2 P_2 - \log_2 P_1}{t_2 - t_1}. \tag{8.5}$$

Since logarithms to the base 10 may be converted to logarithms to the base 2 by multiplying by 3·322, we may substitute $3{\cdot}322r_u$ for r_c in eq. (8.3) to obtain the generation time (g) in terms of common logarithms:

$$3{\cdot}322 r_u = \frac{3{\cdot}322}{g} = \frac{\log_{10} P_2 - \log_{10} P_1}{t_2 - t_1},$$

therefore

$$g = \frac{3{\cdot}322(t_2 - t_1)}{\log_{10} P_2 - \log_{10} P_1}, \tag{8.6}$$

The mean generation time (g) or the time taken for the population to double its mass may vary from about 15 min for some bacteria to several days for certain filamentous fungi.

It is important to note that doubling of cellular matter in unicellular micro-organisms depends mainly upon rate of cell division, but that the same effect can be obtained in filamentous organisms by production of new growing points as lateral branches.

The total growth achieved is determined experimentally. The lag period may be determined graphically by extrapolation to the time axis of the exponential portion of the growth curve. Alternatively, the lag period may be calculated as the difference between the time found to be required for accumulation of a given protoplasmic mass (or its equivalent) and the calculated time needed to achieve the same amount of growth on the assumption that growth is exponential from the beginning.

1.1c. *The stationary phase.* Depletion of nutrients in the medium or accumulation of waste products will eventually limit the exponential increase in cellular matter: the less favourable environmental conditions then cause progressive decrements in the rate of growth until the stationary phase is reached, when the amount of cellular matter in the culture remains constant.

In unicellular organisms the change to the stationary phase may be completed quickly, but in filamentous organisms there is often a rather prolonged period of more limited growth, during which dry weight may continue to increase substantially but in an approximately linear manner. This may be due to gradual diminution in the rate of production of new branched filaments along with continued apical growth of existing hyphae; alternatively, it may be due to deposition of food reserves or the formation of antibiotics and other metabolites by metabolic shunt mechanisms. This intervening period between the logarithmic and stationary phases is sometimes termed the retardation phase.

In summary, we may conclude that the growth curve for any given micro-organism when plotted on a convenient logarithmic scale may be described by the lag period, by the rate of growth and by the total growth achieved.

1.2. Manufacture of commercial bakers' yeast

Special strains of *Saccharomyces cerevisiae* are selected for use in the baking industry. The properties required are consistent ability to produce carbon dioxide rapidly in the flour dough (this requires a high respiration rate, but a low rate of alcoholic fermentation), ability to disperse readily in water and good keeping qualities through ability to resist autolysis. Mild flavour, good appearance in bulk and ability to propagate rapidly to give a good yield in media used for production are also desirable.

The essential process for manufacture of commercial yeast consists in growing a mass culture of the selected pure strain under highly aerobic conditions optimum for logarithmic growth, while limiting the sugar concentration to a low level, thereby minimizing the tendency to produce ethyl alcohol through operation of the Pasteur effect (see Chapter 9).

It is usual to begin a batch fermentation by transfer of a single isolated cell of the preferred strain of *S. cerevisiae* to a nutrient agar slope and to use this culture, when well grown, as inoculum for a static flask culture in liquid medium containing about 6% glucose. Successive transfers into progressively larger vessels containing similar liquid media are effected at intervals of 12–24 hr, with gradually increased aeration, until enough cells have been produced to permit up to about a 20% inoculum into the final production stage having a capacity of 10,000–50,000 gal. By arranging that transfer of the inoculum, at least to the last production stage, is made when the cells are in the logarithmic phase of growth, it can be ensured that no lag phase occurs in the final medium and that the whole culture will continue to grow logarithmically. The production time is determined by the nature of the medium, the rate of nutrient supply and the quantity and physiological status of the inoculum, temperature, aeration and agitation.

It is usual to begin the fermentation at 25°C and to permit a gradual increase up to a maximum of 30°C at the end. Adequate aeration is one of the most important factors in yeast production, and it must be gradually increased to keep pace with the increased growth of the culture. It is necessary to control the pH to between pH 3·5–4·5 throughout to maintain optimum conditions for growth: the low pH reduces contamination by wild bacteria, but if it is allowed to fall too low the colour of the yeast may be adversely affected.

Various carbohydrate and nitrogen sources are employed (see Chapter 6), but it is important to note that *S. cerevisiae* strains can utilize only hexose sugars and that saccharification of the carbohydrate source may therefore be necessary. Some form of phosphorus must be added, and frequently the medium is supplemented with magnesium sulphate. Biotin, pantothenate and inositol are required for maximum yield, and it is necessary to provide these by a suitable blend of the complex sources of accessory food factors (see Chapter 6).

The medium ingredients are usually supplied at a predetermined and often logarithmically increasing rate throughout the production period until the supply is terminated about 1 hr before harvest. This last hour of production proceeds under diminished aeration and is claimed to have a beneficial effect on the product's keeping qualities.

On completion of the production period, the yeast will have increased in weight to approximately five times that of the inoculum, and the cells are harvested by rapid cooling, centrifugation and washing to give a yeast-cream that is passed through filter presses and emerges as a solid containing about 25% dry weight. This is blended, mixed with a plasticizer and extruded in blocks that are cut, wrapped and packed in cartons before cold storage.

1.3. *Production of food and fodder yeast*

Yeasts of both types are produced in a similar manner, but food yeast is usually handled and supplied as a more refined product. Details of manufacture are similar to those already described for commercial bakers' yeast, as might be expected, since the objective is again to produce the maximum number of cells in as short a period of time as possible. Food and fodder yeast provide high protein products that are also a rich source of the vitamin B complex.

Food yeast, and more especially fodder yeast, production permits more variety in the substrates that can be used than does commercial yeast manufacture. This is because yeasts other than *S. cerevisiae* are acceptable, and these strains are not limited to utilization of hexoses as their carbon source. The asporogenous yeast, *Candida utilis* (*Torulopsis utilis*), known as "Torula yeast", is commonly used for these purposes, but *Candida arborea*, *C. pulcherrima*, *C. reukaufii*, *C. tropicalis*, *Hansenula anomala*, *H. suaveolens*, *Saccharomyces cerevisiae* and *Trichosporon pullulans* have also proved useful. Desirable characteristics of a food or fodder yeast are high nutrient value, agreeable flavour, stability in culture, capacity for high growth rate and yield, ability to assimilate diverse sources of carbon and nitrogen, good appearance and ease of processing after harvest.

Candida utilis because of its marked ability to assimilate various carbon sources, including galactose, xylose and a wide range of organic acids, has been used for producing food and fodder from the acid hydrolysates of wood and other cellulosic materials, from citrus waste, from olive residue hydrolysates and from the waste protein water of potato starch manufacturing plants.

1.4. *Mass culture of algae*

Green algae, and in particular the unicellular forms, including species of *Chlorella* and of *Scenedesmus*, produce protein of high nutritive value for animals. They have therefore been studied extensively as sources of protein and vitamins for supplementing the food of undernourished populations.

Unicellular green algae are plants of microscopic size. Their growth requirements of favourable temperature, supply of mineral nutrients and adequate aeration are, in principle, similar to those of yeasts.

Although capable of utilizing sugar as a carbon source, the algae differ in that, given adequate illumination, they are able to meet their carbohydrate requirements by fixation of carbon dioxide. It is this capacity for photosynthesis that has encouraged study of their potential economic uses.

Indoor culture by artificial illumination is uneconomic, but it has been studied as a technique for helping to understand algal growth. Inoculum for a batch fermentation is built up in much the same way as for food yeast production, except that it is usual to begin with an agar slope culture of the selected algal strain rather than a single-cell isolate from a particular strain. Desirable characters in suitable algal strains are high growth rate, high nutritive value, resistance to contaminating micro-organisms, low foam production during growth, cells of suitable average size (the cells must not be so large that they sediment readily but must not be so small that they present difficulty in separation from the medium at harvest) and, for use in tropical countries, strains having high temperature tolerance (some chlorella strains will withstand a temperature of 42°C). The culture on agar containing about 3% w/v glucose as carbon source is inoculated directly into the growth medium contained in a conical flask aerated by agitation on a rotary shaker. This culture is transferred in the logarithmic growth phase into similar medium, but devoid of carbohydrate, and the mixture is stirred, aerated and illuminated. A temperature of 25°C is optimum for most species of *Chlorella* and of *Scenedesmus* at pH 6–6·8, with continuous artificial light intensity of the order of 50–75 kilolux (1 kilolux of 400 to 700 mμ radiation is approximately equivalent to energy of 0·0062 cal/cm^2/min) provided by fluorescent or neon tubes contained in water-tight pockets within the culture solution. The air that is bubbled through the culture is usually supplemented with about 5% v/v CO_2 to prevent limitation of the rate of photosynthesis. Growth is rapid under these conditions, and the culture can be maintained continuously in logarithmic growth, producing cells of high nitrogen content, if successive batches of fresh medium under similar environmental conditions are inoculated with cells

(0·02–0·1 g/l.) transferred from logarithmically growing cultures. Alternatively, the culture can be maintained as a constant population by utilizing steady-state culture techniques (see Chapter 19).

Algae are more critical than yeasts in their need for microelements, but this difficulty has been overcome by providing iron, magnesium and other elements in readily available forms as the metal chelates of ethylenediaminetetracetic acid (EDTA).

Under optimum carbon dioxide and nutrient conditions, the growth rate is a function of temperature and the light intensity available to the individual cells in the algal culture. When the population density is low, so that mutual shading of the algal cells is negligible, growth is exponential (a function of the quantity of algal cellular matter in the culture); when mutual shading results in total absorption of the incident light by the cell suspension, growth becomes linear (a function of the area of the illuminated surface of the culture).

Mass culture of algae as an economic process must be limited to outdoor cultures that can utilize solar energy. High population density of algal cells must be maintained so as to absorb completely the incident light and reduce the cost of separation of the algal cells from the medium at harvest. As we have already seen, high population density results in linear growth rate of the culture when increase in cellular matter is proportional to the area of the illuminated surface: outdoor mass cultures must therefore provide a large ratio of illuminated surface to volume, in which the depth of the illuminated culture solution is limited to between 5 and 20 cm. The population density and composition of the medium are maintained roughly constant at about 2–4 g/l. by harvesting the algal cells in linear growth at intervals when the medium is replenished.

Continuous aeration is necessary, but the supplementary supply of carbon dioxide can be confined to the period of daylight illumination when photosynthesis is active. It is important to note that, when combustion of coal is used to provide carbon dioxide, precautions must be taken to remove sulphur dioxide and hydrogen sulphide before the gas train enters the culture.

Contamination of outdoor cultures by bacteria, fungi, blue-green algae and, more especially, protozoa and rotifers can be deleterious. Both 2,4-dinitro-6-cyclohexylphenylacetate and pentachlorophenylacetate (1–3 μg/ml) have been used to kill the protozoa and rotifers without effect on the desired algae.

Yields of 7–20 g/m^2/day, depending on the type of apparatus, appear to be feasible in outdoor cultures with favourable light and temperature; the product is normally prepared as a dried powder for supplementation of an inadequate diet.

1.5. *Microbial fat*

Filamentous fungi, yeasts, algae and some bacteria are capable of producing substantial amounts of lipid material when grown under appropriate conditions. A small proportion of this lipid potential is represented by material that forms part of the structure of the microbial cells, but most of it is present as food-storage materials.

The fatty-acid content of the fats of micro-organisms bears a general resemblance to that of plant and animal depot fats and has therefore been considered to be potentially suitable for human and animal consumption. The economics of production of extracted microbial fat are, however, unfavourable, but this could become more favourable if particular "essential unsaturated acids" (linoleic and linolenic acids) gained economic importance. Nevertheless, micro-organisms, and filamentous moulds, in particular, are of considerable potential value because the whole dried mycelium, of high fat content and suitable for animal feeds, can be produced on low cost waste liquors. In these processes, growth and fat production by the mould simultaneously lower the biological oxygen demand (see Chapter 16) of the waste liquor, thereby rendering the effluent more suitable for disposal.

Numerous genera of filamentous fungi are worth consideration for production of microbial fat: these include *Penicillium*, *Aspergillus*, *Fusarium*, *Oospora*, *Rhizopus*, *Phycomyces* and *Mucor*. Yields of fat per dry weight of mycelium are of the order of

25–50% under favourable conditions. *Aspergillus nidulans*, *A. flavipes*, *Penicillium javanicum*, *P. soppi* and *Fusarium lini* have been extensively investigated, with particularly favourable results.

Production of microbial fat is essentially a two-stage process: first, the mould mycelium must be produced in bulk and then it must be induced to deposit fat. Conditions for the first stage are those that favour logarithmic growth: an adequate supply of all essential nutrients, favourable pH (about 4 to 5·5 for filamentous fungi), good aeration and a relatively low carbon : nitrogen ratio. Towards the end of the logarithmic phase of growth the nitrogen concentration should be allowed to decrease while the carbohydrate supply is being increased and good aeration is maintained. Optimum conditions for fat production are attained when aeration is good, the carbon:nitrogen ratio is high and magnesium and phosphorus are supplied at critical concentrations. With fixed nitrogen concentration, fat production by *Aspergillus niger* increases with accession of carbohydrate up to 20% w/v concentration; conversely, with fixed glucose concentration, fat production decreases as the nitrogen concentration is increased. Glucose, and xylose when it can be metabolized, have been found to be among the best carbohydrate sources for fat production.

Fat production may continue for a considerable time and will be reflected in the growth curve (as measured by dry weight of mycelium) by prolongation of the "retardation phase" (Fig. 8.1).

The component fatty acids and the composition of the glycerides of microbial fat depends to a large extent upon the micro-organism, but environmental conditions can cause some modifications. In *A. nidulans* the mixed fatty acids become more unsaturated with rise in pH, mainly due to increased content of oleic acid, and maximum unsaturation with higher proportions of oleic, linoleic and linolenic acid occurs when the temperature is lowered to about 17°C.

1.6. *Production of mushroom mycelium*

Higher fungi belonging to the basidiomycetes include many

species that are considered by gourmets to be delicacies; a few, such as the mushroom (*Agaricus bisporus* derived from the wild *A. campestris*) and the morel (*Morchella* spp.) are valuable economic products. It is the sporophore of the fungus (collected from its natural habitat or grown in artificial composts) that has, until recently, constituted the edible delicacy; with the development of canned soups, attempts have been made to produce an acceptable substitute by growing the mycelium in submerged culture.

Agaricus bisporus and *Morchella* spp. can be readily grown in surface cultures on agar media of a wide variety, since they can both utilize a wide range of carbohydrate and nitrogen sources for growth. Both can also be grown in submerged culture, but some strains show much greater tendency to produce pellets of growth rather than a homogeneous broth. The tendency to pelletted growth in submerged culture is often exhibited by an organism that is being grown under sub-optimal conditions: the condition can sometimes be rectified by use of larger inoculum, higher aeration, change in initial pH or inclusion of accessory food factors in the medium. Sometimes pelletted initial growth will give place, after a variable period, to a homogeneous and vigorously growing mycelium without any adjustment to the controlled environmental conditions.

Reports to date indicate that *Morchella* mycelium, in the form of pellets, can be grown in large-scale submerged culture and that the dried mycelium retains the characteristic flavours of the natural sporophores, although individual strains have shown much variation. *Agaricus bisporus* can be grown well in submerged culture as a homogeneous mycelium, but this has lacked, up to now, the true mushroom flavour.

It is important to note that the sporophores consumed as morels and mushrooms are formed from a secondary mycelium of binucleate cells arising from anastomoses between the uninucleate or primary mycelium originally produced on germination of the basidiospores. Whether or not flavour in the submerged culture of mushrooms and morels is related to formation of the binucleate mycelium will presumably be settled with time, but it is clear that this morphological and perhaps physiological feature involves an additional complication that needs to be considered when devising methods of production.

2. Micro-organisms as accessory agents to food production

2.1. *Enzymes*

Industrial microbiological processes depend upon the co-ordinated action of large numbers of enzymes produced by the viable micro-organisms. In recent years the specific enzymic capacity of various micro-organisms has been used to effect chemical changes in substitution for less efficient chemical processes (see Chapter 17). A natural development was to consider possible uses of individual enzymes rather than the whole living micro-organism for effecting particular chemical transformations. This approach is advancing rapidly, and it seems probable that in due course microbial enzymes may be substituted for micro-organisms in a variety of industrial processes. Meanwhile, the first industrial enzymes are being utilized in the food industry. Extraction and purification of enzymes is costly, so that at present the commercial products are relatively crude, although precautions are usually taken to minimize the activity in the product of other enzymes that are not required.

2.1a. *Biosynthesis of enzymes.* We have seen in Chapter 3 that enzymes are generated in strict accordance with the genotypic information contained within the chromosomes, but the precise details of processes regulating the synthesis and function of enzymes are not known.

The natural amino acids are activated by a supply of chemical energy from adenosine triphosphate (see Chapter 9) under the influence of relatively specific enzymes, with formation of amino acyl adenylates containing energy-rich bonds. The activated amino acids are stabilized by their formative enzymes, and each amino acyl moiety is then transferred directly to its specific "transfer ribonucleic acid" (sRNA) (see Chapter 3): the molecules of all forms of sRNA are believed to contain a particular pentanucleotide sequence as a common feature. The carboxyl group of the activated amino acid is transferred directly to an hydroxyl group of ribose in the terminal adenosine nucleoside with formation of an ester that conserves enough energy for subsequent formation of peptide bonds.

Enzymes, in common with other proteins, may be synthesized in many different parts of the cell, but always in association with cytoplasmic particles of ribonucleoprotein structure known as ribosomes, so that a correlation exists between the amount of ribonucleic acid (RNA) in a cell and its capacity for making protein. Ribosomes ensure correct alignment of the activated amino acid residues in correspondence with the specific sequence of bases in the genetic nucleic acid before formation of the peptide bonds that generate the polypeptide chains comprising the enzyme.

It seems likely that a specific sequence of possibly three nucleotides in the nuclear DNA polymer acts as a code for any one particular amino acid in the polypeptide chains of the enzyme and that, although the coding units are contiguous, they do not overlap. This genotypic information is probably transferred by a template mechanism to "messenger RNA" (mRNA), which then carries the instructions to the ribosomes of the cytoplasm as described above.

Ribosomes occur in clusters of predominantly five units in linear order, linked by RNA: these clusters are termed polyribosomes or polysomes. It seems likely that the RNA of the polysome is mRNA and that oncoming ribosomes without a polypeptide chain become attached at one end of the strand of mRNA and then gradually move along, like raindrops on a telephone wire, synthesizing by proximal growth a polypeptide chain from activated amino acid residues provided by the sRNA as they progress. The polypeptide chain appears to be completed when the ribosome reaches the other end of the mRNA strand. The ribosome and its completed polypeptide chain are then released from the polysome before the polypeptide chain is freed. Spontaneous folding and stabilization of the polypeptide chain by juxtaposition of a few key amino acids must follow under suitable environmental conditions (pH, ionic concentration, temperature), sometimes accompanied by association of polypeptides and addition of prosthetic groups ("active groups" tightly bound to enzyme protein), in order to complete the fully functional enzyme. Occasionally, activation of a precursor protein is necessary before the enzyme becomes functional. Spontaneous associations between completed enzymes or between enzymes and other proteins may then form multi-enzyme complexes having an ordered spatial arrangement and capable of catalysing a sequence of chemical reactions.

2.1b. *Factors affecting the amounts of enzyme synthesized.* Although the potential for production of different kinds of enzymes by a micro-organism is determined by its genetic composition, the amounts in fact produced may be greatly influenced by the nutritional and physical conditions of the environment.

Enzymes may be classified into three types, depending upon the effect of environment on the quantities produced. Constitutive enzymes are formed in relatively constant amounts little affected by environmental conditions; inducible enzymes require the presence of an inducer compound before they are formed other than in small amounts; repressible enzymes appear in large amounts but only when the appropriate repressor compounds are absent. Despite these convenient categories, the distinction between constitutive and weakly inducible enzymes is sometimes marginal, and both appear to be formed by a similar mechanism.

Enzyme induction is due to physiological change in the whole microbial population, and it involves an accelerated rate of renewed formation of the enzyme in response to a relatively specific nutritional stimulus: sometimes production of an inducible enzyme can continue only so long as the inducer remains present.

Formation of repressible enzymes is not simply due to activation of a precursor by removal of the repressor compound: protein synthesis is necessary for formation of these enzymes.

Constitutive, inducible and repressible enzymes are all expressions of the genotype but the determinant mechanisms are not understood. At the present time a gene is believed to determine the structure of the enzyme, a repressor to prevent expression of the gene and an inducer to release the gene from inhibition by the repressor.

2.1c. *Amylases.* α- and β-amylases are recognized, the former degrading starch or glycogen to dextrins and the latter producing predominantly maltose from the same substrates. Fungal amylases are used extensively in food industries. Selected strains of *Aspergillus oryzae*, *A. niger*, *Rhizopus delemar*, *R. oryzae* and *Mucor* spp. are employed for enzyme production under varied conditions permitting good growth, and the enzyme is extracted from the broth when growth is almost completed. Commercial amylases are usually marketed as partly purified products.

Bacillus subtilis has also been used for production of amylases, but most strains produce mainly α-amylase.

2.1d. *Miscellaneous enzymes.* Glucose oxidase from *Aspergillus niger* or *Penicillium notatum* oxidises D-glucose to gluconic acid: it is used in food industries to improve the shelf-life of canned soft drinks and to free the white of eggs from glucose, thereby aiding drying with concomitant improvement in stability and subsequent reconstitution.

Microbial proteases are employed for proteolysis of certain fish products, for tenderization of meat and for clarification of some malt beverages.

A mixture of pectolytic enzymes collectively known as pectinases is employed for clarification of fruit juices, wines and vinegars by hydrolysis of pectin. It is usually produced from species of *Aspergillus* and *penicillium* grown in the presence of pectin or related compounds.

Considerable developments in the production and utilization of industrial enzymes are likely over the next few years, and the optimum conditions for their production may be expected to become clarified.

2.2. *Bread*

Enzymes play an essential part in the production of bread. A relatively large inoculum of a selected strain of the yeast, *Saccharomyces cerevisiae*, is added to a mixture of flour and water known as dough, and this is sometimes supplemented by addition of dough conditioners (ammonium salts as medium ingredients for the yeast and other salts to improve dough quality). Fermentation begins at once and continues, usually at a temperature of about 27°C, until the yeast enzymes are inactivated by baking. Under these relatively anaerobic conditions there is little growth, but the dough is changed both chemically and physically. The natural amylase of flour, sometimes supplemented by addition of fungal amylase, hydrolyses the starch to sugars, and these, supplemented by addition of more sugar or malt, are fermented, mainly to ethanol and carbon dioxide, the liberated gas leavening or raising the dough to give the desired porous texture. The ethanol, together with some minor fermentation products (acids, esters and aldehydes), may contribute to the flavour of bread, but much of the characteristic flavour associated with home-made bread may be due to metabolic products formed by contaminating bacteria that are able to grow during the relatively longer periods allowed for leavening and working.

Acid production during fermentation of the carbohydrate is sometimes supplemented by additives to lower the pH, while proteolytic enzymes (from flour, yeast or added fungal enzymes) attack the flour proteins (gluten). This maturation process induces an elastic and springy texture in the dough, thereby aiding retention of a maximum amount of carbon dioxide produced during fermentation.

The leavening process is controlled by adjustment of the yeast inoculum, the sugar concentration, the temperature and the concentration of added salts, to provide dough of requisite texture and able to hold sufficient gas at the right time.

The baking process kills the yeasts, inactivates the enzymes, expands the included gas, produces a desirable appearance to the loaf, generates desirable flavours, confers structural strength to the loaf, renders the residual starch edible by gelatinization and eliminates most of the ethanol and other volatile metabolic yeast products.

3. Microbial spoilage of food

Most foods are potentially good substrates for growth of micro-organisms, and some foods depend for their quality on the microbial growth within them. Food spoilage occurs when the wrong microbes are able to grow or when growth of the desired micro-organism is excessive. Contaminants may damage food by development of off-flavours or by unwanted changes in physical appearance or texture, or they may cause gastrointestinal upset in the consumer or more serious disease because of their pathogenicity.

Methods of food preservation are designed to exclude micro-organisms (asepsis); to remove them (filtration); to kill them before they can do harm (sterilization); or to render the environment unsuitable for their growth (inclusion of preservatives, low temperature storage, anaerobic packs, unavailability of water after drying, high osmotic concentration or inclusion of hydrophilic colloids). The principles of the methods underlying food preservation have been considered in Chapter 5.

The particular spoilage observed will depend upon the kinds of contaminating species, their mutual compatibility, the size of inoculum and the influence of ecological factors upon the association of micro-organisms. These factors are the chemical composition of the food in relation to availability of nutrients and accessory food factors, the temperature of storage, oxygen supply, moisture content, pH and buffering capacity, oxidation-reduction potential, osmotic concentration and presence or absence of inhibitory substances.

Spoilage of alcoholic beverages is considered in Chapter 10 and spoilage of vinegar in Chapter 17.

3.1. *Cereal grains and meals*

The source of spoilage organisms is the natural flora of the grains, and spoilage may be initiated by high moisture content of the seeds. Various fungi, including species of *Penicillium* and *Aspergillus*, may be expected to grow on the surface of moistened grain. A moistened mash of cereals may show a microbiological succession. Lactic acid bacteria and coliform bacteria first produce an acid fermentation and then, at a suitably acidic pH, occurs alcoholic fermentation by yeasts. Finally, fungi and yeasts may invade the surface until inhibited by the metabolic products of acetic acid bacteria (see Chapter 17) growing in the bulk food and oxidizing the alcohol to acetic acid.

3.2. *Bread*

Baking temperatures are sufficiently high to kill any fungi within or on the surface of the loaf, but spoilage may occur by contamination during cooling, handling, slicing or wrapping: spoilage may be aggravated by condensation within the pack or by storage in warm damp conditions.

"Mould" in bread may be due to a variety of fungi, including *Rhizopus nigricans* (white, cotton-like mycelium), *Neurospora sitophila* (pink conidia), *Penicillium expansum* (green conidia) and *Aspergillus niger* (black conidia).

"Rope" in bread, observed by its unpleasant odour accompanied by discoloration, and finally by a sticky consistency that enables the loaf to be stretched into strings when it is broken, results when a mucoid variant of *Bacillus subtilis* contaminates the dough or the ingredients employed in its preparation. These bacterial spores remain viable throughout the baking process and subsequently germinate if the loaf is subjected to a warm humid environment. The ropiness is due partly to the capsular material surrounding the bacterial cells and partly to enzymatic hydrolysis of the proteins and starch in the flour.

3.3. *Fruit and vegetables*

The source of spoilage micro-organisms may be infection with plant pathogens before harvest, or there may be subsequent contamination with air-borne spores of saprophytic micro-organisms. The environmental conditions favouring spoilage may be mechanical damage, pre-harvest attack by plant pathogens, and their subsequent development in transit or storage, or improper storage.

Numerous forms of spoilage may occur but the commonest include bacterial soft rot (*Erwinia carotovora*), Rhizopus soft rot (*Rhizopus* spp.), brown rot (*Sclerotinia* spp.), blue mould rot (*Penicillium* spp.), grey mould (*Botrytis cinerea*) and rots due to species of *Gloeosporium:* other rots may sometimes be caused by species of *Alternaria*, *Cladosporium*, *Colletotrichum*, *Ceratostomella*, *Diplodia*, *Fusarium*, *Physalospora* or *Phomopsis*. Downy mildews are usually caused by species of *Phytophthora* and *Bremia.*

3.4. *Sugar-containing products*

The resistance of sugar-containing products to microbial attack is due to their high osmotic concentration. Spoilage of these products is therefore confined to air-borne osmophilic micro-organisms: these include some bacteria of the genera *Bacillus* and *Leuconostoc* and particular fungi, including some yeasts.

3.5. *Products from fermented vegetables*

These products include olives, pickles and *Sauerkraut:* the preservation of all depends largely on the organic acids (mainly lactic) produced in their preparation by fermentation processes (see Chapter 10). Spoilage may be due to abnormal fermentation conditions during preparation of the product or to subsequent oxidation of the organic acids by contaminating air-borne moulds (including some yeasts) that are salt- or acid-tolerant. Removal of the organic acids then permits further attack by miscellaneous micro-organisms, resulting in an undesirable appearance, odour or texture of the product.

3.6. *Meat products*

These products are subject to contamination by divers micro-organisms. Invasion of animal tissues after death occurs mainly by those micro-organisms present in the intestinal tract, but serious contamination can also arise from microbes on the exterior of the animal, by introduction from process workers, utensils or equipment used in slaughtering, cutting and dressing, by transfer from vehicles or containers needed for transportation or by air-borne infection at any stage before consumption.

Micro-organisms spread in fresh meat through the blood and lymph vessels and more slowly through the intercellular spaces of connective tissue: any mincing of meat may also facilitate dissemination of the contaminants. The physical properties of the meat, the efficiency of the methods employed to reduce the inoculum potential before slaughter, the cleanliness of processing methods and the rapidity of cooling will together largely determine the degree of contamination. Temperature and the availability of oxygen will largely determine the species of micro-organisms involved.

Meat is a complete medium at a suitable pH for the growth of many micro-organisms: autolytic changes, including partial proteolysis of proteins of muscle and connective tissues and some hydrolysis of fats after death, render it even more suitable. Common forms of microbial spoilage of meat products are listed in Table 8.1.

TABLE 8.1. MICROBIAL SPOILAGE OF MEAT PRODUCTS

Storage conditions	Type of spoilage	Micro-organisms responsible for spoilage
Aerobic	Surface slime	Species of *Achromobacter*,* *Bacillus*, *Leuconostoc*, *Micrococcus*,* *Streptococcus*, and sometimes *Lactobacillus* and yeasts.
	Oxidation of meat pigments producing colour (brown, green or grey)	Miscellaneous bacteria including some that cause spoilage under refrigeration
	Stickiness	Early stages of growth of miscellaneous fungi.
	Localized patches showing mould spores	*Penicillium expansum*, *P. asperulum*, *P. oxalicum* (green); *Cladosporium herbarum* (black); *Sporotrichum carnis* (white).
	Spots of surface coloration	Pigmented bacteria; *Chromobacterium lividum*, *Micrococcus* spp., *Flavobacterium** spp. (yellow); *Pseudomonas** *syncyanea* (blue); *Serratia marcescens* (red).
	Surface phosphorescence	Luminescent species of photobacterium.
	Undesirable odour or taste	Miscellaneous fungi (musty flavour); miscellaneous bacteria ("taints").
	Rancidity of fat	Unsaturated animal fats have a tendency to autoxidation in air; the process may be accelerated by light or by contamination with yeasts, other fungi and lipolytic bacteria of the genera *Achromobacter* and *Pseudomonas*.

TABLE 8.1.—*Continued*

Storage conditions	Type of spoilage	Micro-organisms responsible for spoilage
Anaerobic	Souring	Production of fatty acids or of lactic acid by anaerobic bacteria; proteolysis by facultative or anaerobic bacteria.
	Putrefaction	Anaerobic decomposition of protein to yield malodorous compounds (amines, ammonia, H_2S, indole, mercaptans and skatole) by *Clostridium* spp. or by facultative anaerobic bacteria (*Achromobacter*, *Pseudomonas* and *Proteus* spp.).

* Some species of these genera cause spoilage even under refrigerated conditions (approx. 0°C).

3.7. *Milk products*

These products are usually only partially sterilized by pasteurization or by tyndallization (see Chapter 5), so that they can be stored only for a limited period.

Milk is an excellent medium for the culture of micro-organisms, and it is therefore susceptible to several forms of spoilage. Souring of milk affects the flavour and eventually results in curdling due to fermentation of the carbohydrate in a manner typical of the lactic acid bacteria (see Chapter 10) with precipitation of protein at the lowered pH.

Lactic acid bacteria are most active in raw milk but some of the thermophilic forms may survive pasteurization. When fermentations by lactic acid bacteria are held in check, surviving spores of coliform bacteria may intervene to produce lactic acid along with considerable amounts of volatile metabolic products; alternatively, spores of *Clostridium* spp. may lead to production of butyric acid accompanied by gaseous hydrogen and carbon dioxide (see Chapter 11).

Proteolysis of milk due to asporogenous but low-temperature-resistant species of *Acromobacter*, *Alcaligenes*, *Micrococcus*, *Flavobacterium*, *Proteus*, *Pseudomonas* and *Serratia* may occur in raw milk chilled for a prolonged period: these forms should be killed by pasteurization but recontamination usually occurs to permit proteolysis if the pasteurized milk is subsequently refrigerated for a relatively long period.

3.8. *Canned foods*

These products are usually subjected to heat treatments of varying degrees of severity during processing, so that most of the spoilage encountered is due to survival of the high-temperature-resistant spores of some thermophilic and facultatively thermophilic bacteria. The only cure is more efficient processing.

Canned foods of low acid content (cereals and legumes) are particularly prone to attack by *Bacillus* spp. producing organic acids but no gas ("flat sour spoilage"), so that the can does not swell. Spoilage due to the thermophilic spore-forming anaerobic *Clostridium thermosaccharolyticum* results in swelling and perhaps eventual bursting of the can when liberation of carbon dioxide and hydrogen gases (but no hydrogen sulphide) accompanies the metabolic conversion of sugars to organic acids (T.A. spoilage).

Some putrefactive anaerobic bacteria of the genus *Clostridium*, although normally killed by high temperature, have spores that are highly resistant to heat, so that putrefaction accompanied by swelling of the container may sometimes occur in canned food. One of these micro-organisms, *C. botulinum*, is sometimes the cause of food poisoning.

4. Infected food and human illness

Contaminated food may serve as a passive carrier for numerous microbial infections: these include brucellosis (*Brucella* spp.), cholera (*Vibrio cholerae*), diphtheria (*Bacillus diphtheriae*), dysentery (*Shigella* spp.), tuberculosis (*Mycobacterium tuberculosis* var. *hominis*) and typhoid fever (*Bacillus typhosus*).

Other contaminating micro-organisms may use the food as a substrate for growth, multiplying to such an extent that a considerable hazard occurs when the food is eventually consumed. This may be due to the presence of potent bacterial toxins (botulism and *Staphylococcus* poisoning) or to the increased probability of infection from the high inoculum potential of pathogenic organisms (*Salmonella* infection). In the United Kingdom, salmonellae are much the most important cause of food-poisoning, but in the U.S.A. *Staphylococcus* poisoning appears to be of greater importance.

4.1. *Botulism*

This form of food-poisoning results from ingestion of a potent proteinaceous and somewhat thermolabile exotoxin, produced in food by growth of various types of *Clostridium botulinum*, an anaerobic soil-inhabiting organism included in the Bacillaceae family of rod-shaped bacteria that produce endospores. These anaerobes putrefy proteinaceous food or ferment carbohydrates, and both processes are accompanied by gas production. Meat and other foods having an obnoxious odour should therefore be rejected, together with any canned foods that exhibit pressure in the container.

Botulism is normally prevented by employing cleanliness to reduce spore contamination and by adequate processing during canning to kill any residual spores. *C. botulinum* spores usually need a temperature above 10–15°C for germination, but growth is optimum at 37°C under not excessively acid conditions: much can therefore be done by taking suitable precautions in food preparation to provide environmental conditions that prevent germination, growth and toxin production after processing or canning.

4.2. Clostridium welchii *poisoning*

Certain strains of *C. welchii*, widely distributed in excreta of human, animal or fly origin, have been reported in recent years to be the cause of outbreaks of food-poisoning, particularly in institutions.

Poisoning due to *C. welchii* is usually relatively mild. Bulky dishes of cooked meats present the main hazard, particularly when maintained before consumption at an elevated temperature that permits anaerobic growth. Personal hygiene and cleanliness in the kitchen helps to prevent infection, but raw meat may be already contaminated when purchased. Attention to the cooking and storage of meat dishes offers the main safeguard. Minimum opportunity for incubation of the contaminant micro-organisms occurs when cuts of meat do not exceed 6 lb in size and when the meat is cooked and eaten at once. When delay in consumption is unavoidable, the cooked meat should be cooled rapidly and refrigerated until required; if it must be served hot, the meat should be maintained at a temperature exceeding 60°C.

Clostridium welchii poisoning may also be diminished by avoiding the re-heating of meat dishes and by ensuring that stock-pot ingredients are maintained at boiling temperature and that the surplus is discarded each day.

4.3. Staphylococcus *poisoning*

Contamination of many kinds of preserved food with toxin-producing strains of *Staphylococcus aureus* from human and animal sources (boils, infected wounds, nasal and sinus infections) is the most commonly encountered form of food poisoning in the U.S.A. These gram-positive cocci of the family Micrococcaceae do not usually render the food obnoxious or unattractive. Personal hygiene in handling of food, heat treatment of all milk supplies and adequate facilities for cold-storage of semi-preserved foods afford the best protection against *Staphylococcus* poisoning.

4.4. Salmonella *infection*

Several species of *Salmonella*, gram-negative motile rod-shaped non-sporing bacteria included in the family Enterobacteriaceae, may contaminate a variety of foods. The micro-organisms occur in the intestinal tract, and infection is spread by consumption of food contaminated with *Salmonella*-infected faeces. Infected animals, especially poultry and their eggs and flies, are the usual source of species (commonly *S. typhimurium*) responsible for human gastroenteritis. Fortunately, *Salmonella* spp. are readily killed by heat, so that pasteurization or adequate cooking combined with personal and animal hygiene helps to reduce the possibility of infection. Considerable growth of the contaminating bacteria within the food is necessary to provide the high inoculum needed for infection: refrigerated storage and avoidance of prolonged periods of food-warming will therefore again offer the best means of protection.

Suggested literature for further reading

1. CHAMPAGNAT, A., VERNET, C., LAINÉ, B. and FILOSA, J., Biosynthesis of protein-vitamin concentrates from petroleum, *Nature, Lond.* **197,** 13–14 (1963).
2. CHANTRENNE, H., Aspects of the biosynthesis of enzymes, *Adv. Enzymol.* **24,** 1–33 (1962).
3. COCKBURN, W. C., TAYLOR, J. ANDERSON, E. S. and HOBBS, B. C., *Food Poisoning (Symposium)*, the Royal Society of Health, London, 1962.
4. FRAZIER, W. C., *Food Microbiology*, McGraw-Hill, New York, Toronto, London, 1958.
5. GOODMAN, H. M. and RICH, A., Mechanism of polyribosome action during protein synthesis, *Nature, Lond.* **199,** 318–22 (1963).
6. GORESLINE, H. E., INGRAM, M., MACÚCH, P., MOCQUOT, G., MOSSEL, D. A. A., NIVEN, Jr. C. F. and THATCHER, F. S., Tentative classification of food irradiation processes with microbiological objectives, *Nature, Lond.* **204,** 237–8 (1964).
7. HALVORSON, H. O., Food microbiology, *A. Rev. Microbiol.* **3,** 357–70 (1949).
8. JEZESKI, J. J., Microbiology of dairy products, *A. Rev. Microbiol.* **8,** 429–48 (1954).
9. MARR, A. G., Enzyme localization in bacteria, *A. Rev. Microbiol.* **14,** 241–60 (1960).
10. NIRENBERG, M. W. and MATTHAEI, J. H., The dependence of cell-free protein synthesis in *Escherichia coli* upon naturally occurring or synthetic polyribonucleotides, *Proc. natn. Acad. Sci. U.S.A.* **47,** 1588–1602 (1961).

11. NISMAN, B. and PELMONT, J. *De novo* protein synthesis *in vitro*, *Prog. Nucleic Acid Research* **3,** 235–97 (1964).
12. NOVELLI, G. D., Protein synthesis in micro-organisms, *A. Rev. Microbiol.* **14,** 65–82 (1960).
13. OSWALD, W. J. and GOLUEKE, C. G., Biological transformation of solar energy, *Adv. appl. Microbiol.* **2,** 223–62 (1960).
14. PARDEE, A. B., *The Synthesis of Enzymes. The Bacteria III: Biosynthesis,* 577–630, Academic Press, New York and London, 1962.
15. QUAYLE, J. R., Metabolism of C_1 compounds in autotrophic and heterotrophic micro-organisms, *A. Rev. Microbiol.* **15,** 119–52 (1961).
16. ROBINSON, R. F. and DAVIDSON, R. S., The large-scale growth of higher fungi, *Adv. appl. Microbiol.* **1,** 261–78 (1959).
17. SIMPSON, M. V., Protein biosynthesis, *A. Rev. Biochem.* **31,** 333–68 (1962).
18. STILLER, M., The path of carbon in photosynthesis, *A. Rev. Pl. Physiol.* **13,** 151–70 (1962).
19. STRAUB, F. B., Formation of the secondary and tertiary structure of enzymes, *Adv. Enzymol.* **26,** 89–114 (1964).
20. TAMIYA, H., Mass culture of algae, *A. Rev. Pl. Physiol.* **8,** 309–34 (1957).
21. THATCHER, F. S., Foods and feeds from fungi, *A. Rev. Microbiol.* **8,** 449–72 (1954).
22. WILSON, G., The public health aspect of food microbiology, *Chemy. Ind.* 1854–9 (1964).
23. WOODBINE, M., Microbial fat: micro-organisms as potential fat producers, *Prog. ind. Microbiol.* **1,** 179–245 (1959).

CHAPTER 9

Fermentation Products Related to Sugars

1. Energy-rich bonds

Most micro-organisms are able to use carbohydrate food reserves as a source of the energy and the building blocks needed for biosynthesis of cellular materials: the energy-rich bond, denoted by the sign $\sim$, plays an important role in both of these biochemical processes.

Any chemical reaction that can take place without external assistance must have an inherent tendency for the free energy (the energy made available during the reaction for performance of useful work) of the system to decrease. The change in free energy (i.e. the change in chemical energy) of the system is related to the change in total heat energy by the equation

$$\Delta F = \Delta H - T.\Delta S,$$

where ΔF is the change in free energy of the system; ΔH is the change in total heat energy of the system; T is the absolute temperature of the reaction; and S is the entropy, a measure of the extent to which the total heat energy of the system is not available for performance of useful work.

Chemical energy is easily converted to heat, with the result that chemical reactions are commonly exothermic, but sometimes the accompanying change of entropy may be so large that, despite the loss of free energy, the reaction is endothermic. The thermodynamic feasibility of a chemical reaction can be ascertained only from the change in free energy of the system: the reaction can proceed unaided if there is a decrease in free energy of the system (i.e. the process is exergonic, making ΔF a negative quantity), but when free energy enters the system the reaction cannot proceed without external aid (i.e. the process is endergonic, making ΔF a positive quantity).

Even though thermodynamic considerations show that a chemical reaction is possible, it can only be accomplished when the molecules are in a reactive state, often due to activation by a catalyst.

Synthetic chemical reactions forming complex compounds from simpler molecules are usually endergonic reactions that may need the use of strong acids or alkalis, high temperature and sometimes high pressure methods that are available to the organic chemist. Biosynthesis, on the other hand, must be accomplished under biological conditions at almost "normal" temperature. The problem is solved in cellular metabolism by trapping some of the chemical energy released by katabolism (chemical degradation of complex substrates to simpler products) in the form of the energy-rich bonds of a small number of compounds that serve as biological energy carriers. These bonds, on hydrolysis, release a large amount of energy, and a chemical group bound in this manner is highly reactive. Much of the free energy associated with the energy-rich bond can ordinarily be transferred with the radical; consequently, biological energy-carriers may also be used in priming reactions that incorporate into one of the reactants enough free energy to make the reaction thermodynamically possible. Enzymes can then activate the substrates, so that the reaction can proceed.

Adenosine triphosphate (ATP) is ordinarily utilized as the energy-carrier in biochemical systems, but other important energy-carriers include acetyl-coenzyme A or active acetate (acetyl$\sim$S.Co$\bar{A}$) and uridine triphosphate (UTP): the structures of these compounds are given in Fig. 9.1. These energy-carriers serve the important function of coupling the predominantly exergonic processes of katabolism to the predominantly endergonic reactions involved in biosynthesis (anabolism).

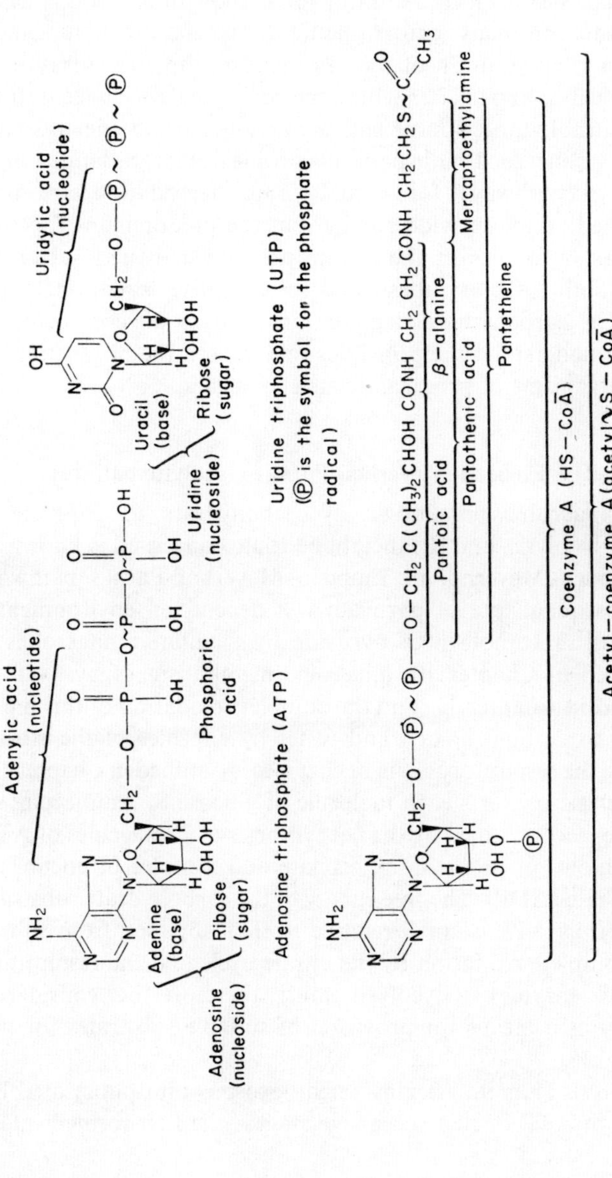

FIG. 9.1. Some important biological energy-carriers.

The manner of degradation of the carbohydrate food reserves will depend on many factors, including the enzymic complement of the particular strain of micro-organism, the concentration of inorganic ions and of coenzymes, the nature and the concentration of the carbohydrate source and the presence or absence of intermediates concerned with particular routes of degradation. Four important pathways for carbohydrate degradation are now recognized: some provide energy, but the predominant function of others is to furnish the coenzymes and activated substrates needed both for biosynthesis and for complete oxidation. These routes for carbohydrate degradation need to be considered in order to understand the principles that govern the production by micro-organisms of products related to sugars.

2. The Embden–Meyerhof–Parnas glycolytic pathway

The anaerobic breakdown of carbohydrate to pyruvate is termed glycolysis, and the metabolic route concerned is known as the Embden–Meyerhof or Embden–Meyerhof–Parnas pathway. The subsequent fate of pyruvate will depend on environmental conditions: metabolism of pyruvate by facultative anaerobes is considered in Chapter 10, anaerobic metabolism of pyruvate is described in Chapter 11, and the complete oxidation of this intermediate to carbon dioxide and water by enzymes of the citrate cycle and the respiratory chain of enzymes is outlined in Chapter 12.

Glycolysis converts each molecule of glucose to 2 molecules of pyruvate, accompanied by the net synthesis of 2 molecules of ATP from 2 moles of inorganic phosphate and 2 moles of adenosine diphosphate (ADP). The free energy of the carbohydrate substrate conserved as ATP is small relative to that obtained from subsequent complete oxidation by the citrate cycle and the respiratory chain of enzymes. Glycolysis must therefore be considered primarily as a means for provision of suitable substrates for the citrate cycle.

Glycolysis (Fig. 9.2) begins when hexose is phosphorylated by 2 moles of ATP in two successive stages: the first-formed inter-

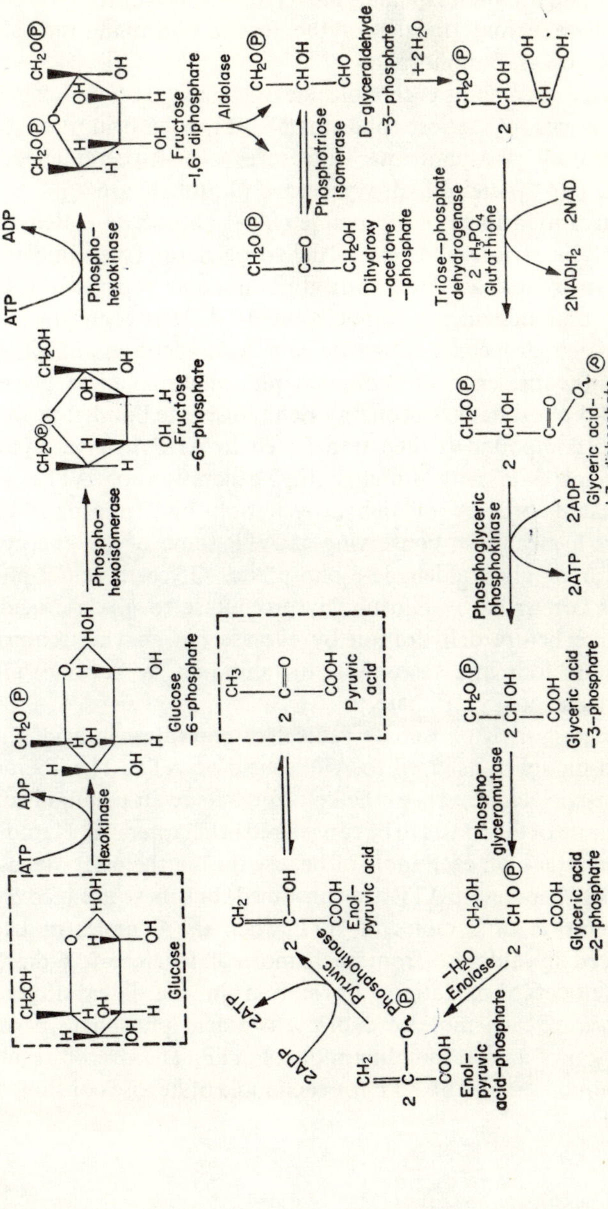

FIG. 9.2. The Embden–Meyerhof–Parnas glycolytic pathway.

mediate, glucose-6-phosphate, is reversibly isomerized to fructose-6-phosphate before transfer of the second phosphate radical to yield fructose-1, 6-diphosphate.

Aldolase then splits each molecule of fructose-1, 6-diphosphate into 2 moles of "triose phosphate", an equilibrium mixture (catalysed by phosphotriose isomerase) of D-glyceraldehyde-3-phosphate (4%) and dihydroxyacetone phosphate (96%).

The hydrated form of glyceraldehyde-3-phosphate is dehydrogenated by triose-phosphate dehydrogenase (susceptible to inhibition by iodoacetate), with glutathione as prosthetic group and nicotinamide adenine dinucleotide (NAD) as coenzyme. The enzyme–dehydrogenase–susbtrate complex undergoes phosphorolysis in the presence of inorganic phosphate to yield glyceric acid-1,3-diphosphate. The energy-rich phosphate bond at position 1 in this compound is then transferred to ADP to give glyceric acid-3-phosphate with simultaneous generation of ATP. This complicated process of dehydrogenation and phosphorolysis provides a means for conserving as ATP some of the energy of oxidation of glyceraldehyde-3-phosphate. Glyceric acid-3-phosphate is converted by phosphoglyceromutase to glyceric acid-2-phosphate before dehydration by enolase (an enzyme requiring magnesium ions and susceptible to inhibition by fluoride) gives enol-pyruvic acid phosphate.

The energy-rich phosphate radical of enol-pyruvic acid phosphate is finally transferred to ADP, yielding ATP, The liberated enol-pyruvic acid reverts to the keto-form to await metabolism by one or more of the routes to be considered in Chapters 10, 11 and 12.

In the attack on each mole of hexose during the early stages of glycolysis, 2 moles of ATP are consumed, but these are recovered by generation of 2 moles of ATP when the 2 moles of triose (produced by aldolase from each mole of fructose-1, 6-diphosphate) undergo oxidation. Formation in the later stages of glycolysis of each mole of enol-pyruvic acid phosphate permits generation of a corresponding mole of ATP. The overall result is a net gain of 2 moles of ATP for each mole of hexose consumed in glycolysis.

It is important to note that the supply of NAD is usually limited, and that glycolysis can therefore continue only so long as provision is made for continuous re-oxidation of the reduced form of nicotinamide adenine dinucleotide ($NADH_2$) that results from dehydrogenation of the glyceraldehyde-3-phosphate–enzyme complex. As we shall see in this and subsequent chapters, the re-oxidation of $NADH_2$ may be effected by several different hydrogen acceptors with concomitant production of various metabolic products.

3. The Warburg–Dickens oxidative pentose-phosphate pathway

Direct and complete oxidation of glucose to carbon dioxide and hydrogen (bound as the reduced form of the coenzyme, nicotinamide adenine dinucleotide phosphate ($NADPH_2$)) by the Warburg–Dickens oxidative pentose-phosphate cycle provides an alternative pathway to glycolysis. The Warburg–Dickens oxidative pentose-phosphate cycle has little significance as an energy source, because the substrates are not transformed in a manner that generates energy-rich phosphate bonds and because re-oxidation of the $NADPH_2$ formed does not appear to be linked to oxidative phosphorylation in the same manner as the re-oxidation of $NADH_2$ (see Chapter 12). On the other hand, $NADPH_2$ appears to be an indispensable reducing agent in biosynthetic systems, and the oxidative pentose-phosphate cycle is its prime source. The Warburg–Dickens oxidative pentose-phosphate cycle is therefore best regarded as an aid to biosynthesis by providing both the specific reducing agent, $NADPH_2$, and the requisitive building blocks, erythrose-4-phosphate, glycolyl residues and ribulose-5-phosphate or its isomerization product, ribose-5-phosphate.

The Warburg–Dickens oxidative pentose-phosphate cycle (Fig. 9.3) begins with dehydrogenation of glucose-6-phosphate to gluconolactone-6-phosphate with concomitant reduction of NADP to $NADPH_2$. Gluconolactone-6-phosphate is readily hydrolysed to gluconate-6-phosphate, the substrate for a second

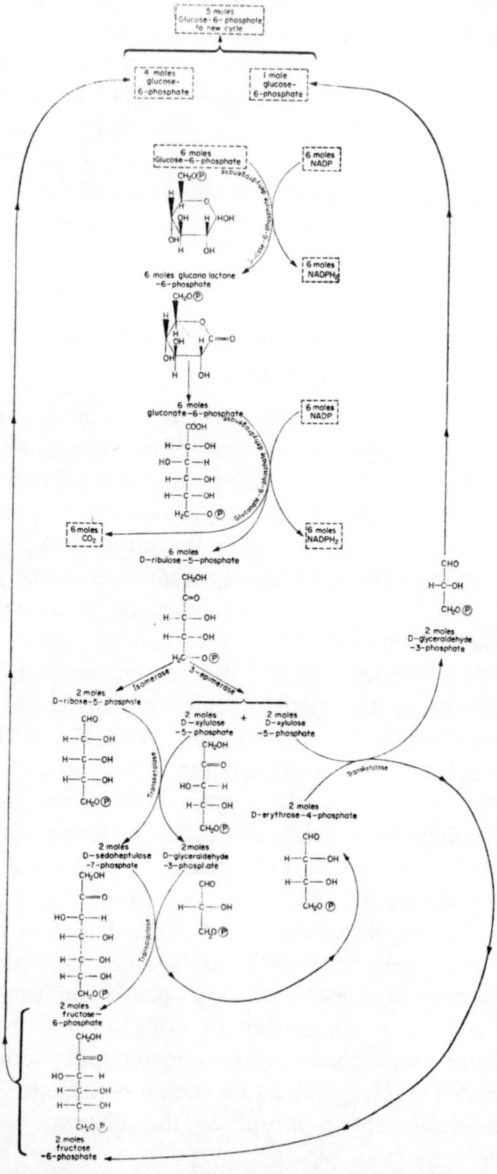

FIG. 9.3. The Warburg–Dickens oxidative pentose-phosphate pathway.

NADP-dependent dehydrogenation that completes the oxidation with simultaneous loss of carbon dioxide to yield ribulose-5-phosphate.

Ribulose-5-phosphate is the substrate for two enzymes—an isomerase yielding ribose-5-phosphate and a 3-epimerase forming xylulose-5-phosphate.

Several transfer and recombination reactions then complete the cycle by resynthesis of glucose-6-phosphate through fructose-6-phosphate and through D-glyceraldehyde-3-phosphate.

The net result is that for every 6 moles glucose-6-phosphate that enter the cycle, 1 mole is completely oxidized to 6 moles carbon dioxide with formation of 12 moles $NADPH_2$.

Since the supply of nicotinamide adenine dinucleotide (NADP) is usually limited, provision must be made for continuous re-oxidation of the reduced coenzyme if the pentose-phosphate cycle is to continue in operation. The re-oxidation may be effected by biosynthetic reactions needing the reduced form of nicotinamide adenine dinucleotide ($NADPH_2$) as specific reducing agent: these include the carboxylation of pyruvate to malate, the reductive amination of α-ketoglutarate to glutamate, described in Chapter 12, and the carboxylation of acetyl-coenzyme A to malonyl-coenzyme A described in Chapter 14. Alternatively, the re-oxidation may be effected by hydrogen transfer from $NADPH_2$ to NAD catalysed by lactate dehydrogenase, an enzyme that can use either of the pyridine nucleotides as coenzyme:

(i)

$$\text{NADPH} + \text{H}^{\oplus} + \underset{\text{pyruvic acid}}{\text{CH}_3\text{CO COOH}} \rightleftharpoons \text{NADP}^{\oplus} + \underset{\text{lactic acid}}{\text{CH}_3\text{CHOH COOH}}$$

(ii)

$$\text{NAD}^{\oplus} + \text{CH}_3\text{CHOH COOH} \rightleftharpoons \text{NADH} + \text{H}^{\oplus} + \text{CH}_3\text{CO COOH}$$

$$\text{Overall: NADPH} + \text{H}^{\oplus} + \text{NAD}^{\oplus} \rightleftharpoons \text{NADP}^{\oplus} + \text{NADH} + \text{H}^{\oplus}$$

The pentoses, ribose and 2-deoxyribose are building blocks for RNA and DNA respectively: it is therefore important to consider the routes available for metabolic synthesis of these sugars.

The Warburg–Dickens oxidative pentose-phosphate cycle forms D-ribose-5-phosphate by oxidation of glucose-6-phosphate with subseqeunt cleavage of C-1 (Fig. 9.3): the cycle can therefore function in the metabolic synthesis of ribose from D-ribose-5-phosphate.

Alternatively, the metabolic synthesis of ribose may occur by reversal of the transaldolase and transketolase reactions shown in Fig. 9.3.

The ribose synthesized by either metabolic route can be reduced to deoxyribose but this reduction occurs at the nucleoside diphosphate level and $NADPH_2$ is specifically required as the reducing agent.

4. Breakdown of glucose by the Entner–Doudoroff pathway

A third route, known as the Entner–Doudoroff pathway, is involved in the metabolism of glucose by some bacteria (e.g. *Pseudomonas* spp.) and by certain fungi, either exclusively or in a supplementary capacity, when either the Embden–Meyerhof glycolytic or the Warburg–Dickens oxidative pentose-phosphate pathway is not fully effective.

The initial stages of the Entner–Doudoroff pathway are the same as for the pentose-phosphate route, but gluconate-6-phosphate is dehydrated by an enzyme requiring ferrous ions, and the enolic product immediately undergoes irreversible tautomerization to give 2-keto-3-deoxy-6-phosphogluconate; this intermediate also is immediately cleaved by aldolase, to give pyruvate and D-glyceraldehyde-3-phosphate.

Hochster and Katznelson have suggested that the Entner–Doudoroff pathway may be followed by a process for resynthesis of hexose phosphates from the mole of D-glyceraldehyde-3-phosphate formed from each mole of glucose consumed. A cyclic process of this nature could ensure that for every 2 moles of glucose entering the cycle, 1 would be completely oxidized via pyruvate and the citrate cycle (see Chapter 12), the other being resynthesized.

5. The glucuronic acid–xylulose cycle

The hexose–pentose cycle given the name glucuronic acid–xylulose cycle (Fig. 9.4) differs from the pentose-phosphate cycle in using mainly unphosphorylated intermediates for degradation of glucose with loss of C-6 instead of C-1.

Each passage of the glucuronic acid–xylulose cycle involves 2 moles of ATP [1 to give UTP, the energy-carrier needed to form the energy-rich compound, uridine diphosphate glucose (UDP-glucose) and the other to form D-xylulose-5-phosphate] and produces 1 mole of $NADH_2$ in the dehydrogenation of UDP-glucose to uridine diphosphate glucuronic acid (UDP-glucuronic acid). The remaining hydrogen transfers produce no overall change in concentration of the pyridine nucleotide coenzymes, because the dehydrogenations and hydrogenations catalysed by either nicotinamide adenine dinucleotide or nicotinamide adenine dinucleotide phosphate are self-compensatory. Continued operation of the cycle will be just as dependent as glycolysis on the regeneration of NAD from the mole of $NADH_2$ produced in each cycle.

The glucuronic acid–xylulose cycle seems quantitatively to be relatively unimportant in the total degradation of carbohydrate, but many of the intermediates play key roles in routes leading to products related to sugars: L-gulonic acid is an intermediate in the formation of L-ascorbic acid, and the nucleoside diphosphate carbohydrates are concerned with the biosynthesis of polysaccharides and polyuronides.

6. Nucleoside diphosphate sugars as intermediates in epimerization of sugars and in biosynthesis of amino sugars, polysaccharides and polyuronides (Fig. 9.5)

Glucosamine-6-phosphate formed from glucose-6-phosphate is the common intermediate for biosynthesis of UDP-glucosamine and its epimer UDP-galactosamine, and for formation of UDP-*N*-acetyl-glucosamine and its epimer UDP-*N*-acetyl-galactosamine:

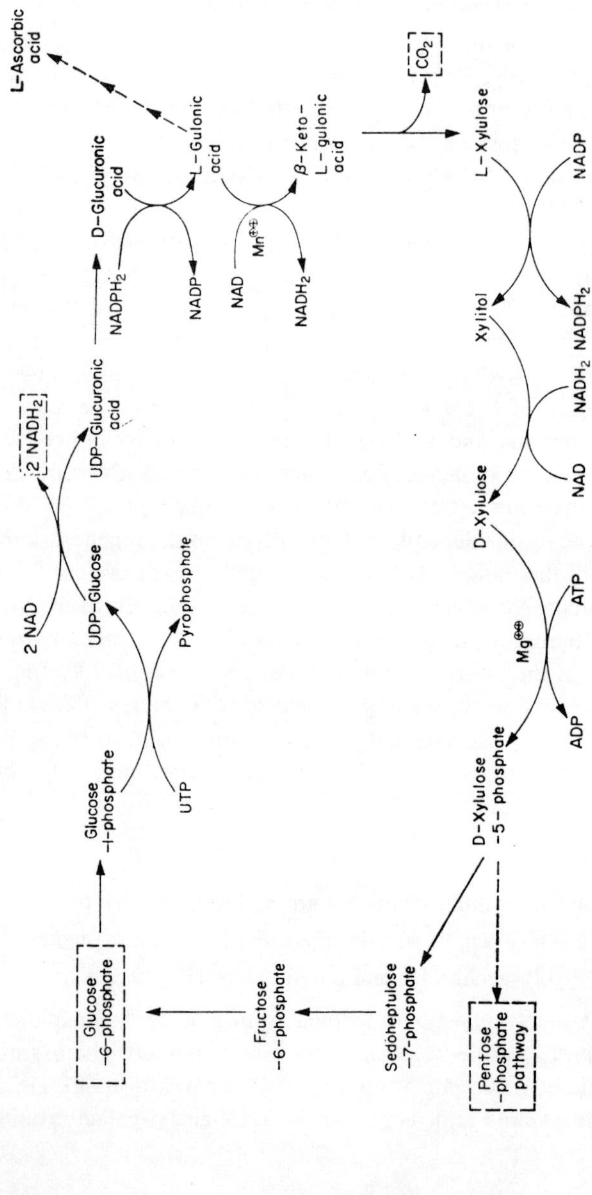

FIG. 9.4. The glucuronic acid–xylulose cycle.

it should be noted that each epimerization takes place at the energy-rich level of a nucleoside diphosphate sugar.

Nucleoside diphosphate sugars, the carriers of active sugars, are also involved in the biosynthesis of polysaccharides, but they now function as glycosyl donors to suitable acceptors in a general reaction, that is, in some cases, reversible:

$$\text{Nucleoside diphosphate sugar} + \text{acceptor} \longrightarrow \text{Nucleoside diphosphate} + \text{acceptor-glycoside}$$

Glucuronides are formed in similar manner by glucuronyl transfer from nucleoside diphosphate glucuronic acid to suitable acceptors.

Some of the more important routes to these sugar-derived products are summarized in Fig. 9.5.

7. Regulation of pathways for breakdown of carbohydrate

Environmental conditions play a large part in determining the metabolic routes employed for breakdown of carbohydrate, and the particular choice of pathway may have a considerable influence on both the yield and type of metabolic products produced by micro-organisms.

7.1. *Regulation of the Warburg–Dickens oxidative pentose-phosphate and Embden–Meyerhof–Parnas glycolytic pathways*

Inorganic ortho-phosphate (Pi), needed for resynthesis of ATP consumed in formation of phosphorylated sugar intermediates (Fig. 9.2), stimulates glycolysis, but has the opposite effect on the oxidative pentose-phosphate pathway, in which 0·05 M inorganic phosphate gives 45% inhibition of glucose-6-phosphate dehydrogenase (Fig. 9.3).

Similarly, ATP needed for biosynthesis of glycolytic route intermediates (Fig. 9.2), inhibits competitively the gluconate-6-phosphate dehydrogenase (Fig. 9.3) of the oxidative pentose-phosphate pathway.

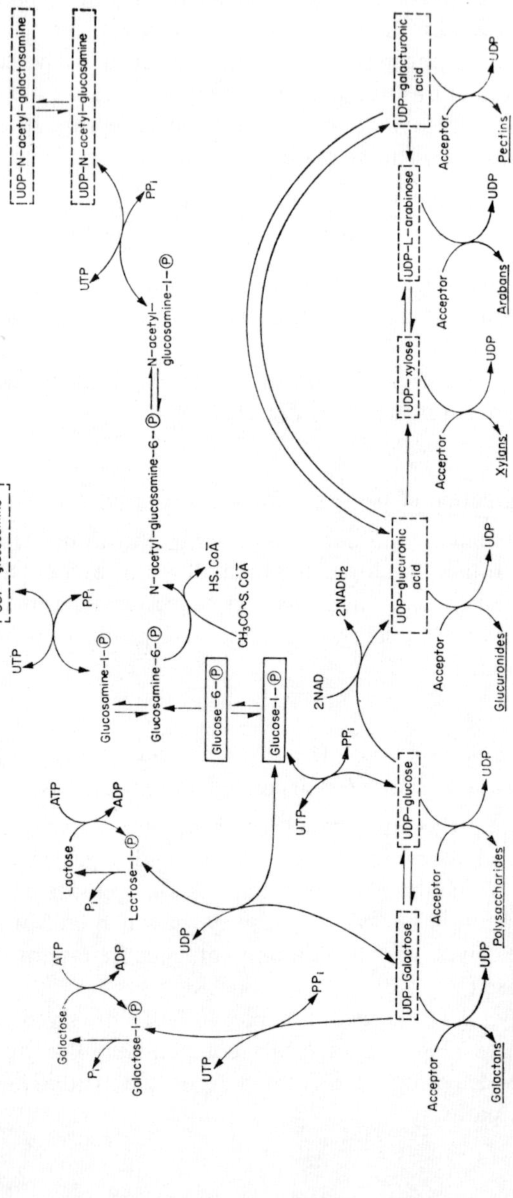

FIG. 9.5. Nucleoside diphosphate sugars as intermediates in biological transformation of carbohydrates.

Most tissues operating the oxidative pentose-phosphate pathway tend towards temporary accumulation of gluconate-6-phosphate, because the activity of glucose-6-phosphate dehydrogenase is usually higher than the activity of gluconate-6-phosphate dehydrogenase (Fig. 9.3). Gluconate-6-phosphate however, is an inhibitor of phospho-hexoisomerase (Fig. 9.2); operation of the oxidative pentose-phosphate pathway therefore usually has an inhibitory effect on glycolysis.

Aeration also has marked but opposite effects on the glycolytic and oxidative pentose-phosphate pathways. In general, good aeration favours the oxidative pentose-phosphate pathway, because glycolysis is then inhibited. Inhibition of glycolysis by aerobiosis is known as the Pasteur effect: the action has been attributed to inhibition by inorganic phosphate deficiency of triose-phosphate dehydrogenase (Fig. 9.2). This deficiency might be expected if oxidative phosphorylation of inorganic phosphate to ATP predominates under good aeration conditions.

The ratio $NAD:NADH_2$ in cells of micro-organisms is usually greater than 1. On the other hand, the ratio $NADP:NADPH_2$ is usually less than 1; moreover, the total content of nicotinamide adenine dinucleotide phosphate coenzyme is low. We have seen that glycolysis depends on continuously available NAD, but that operation of the oxidative pentose-phosphate pathway needs a continuous supply of NADP. These considerations explain the observed markedly stimulatory effect of a small increase in concentration of NADP on breakdown of carbohydrates by way of the Warburg–Dickens oxidative pentose-phosphate pathway. Any factor that increases the rate of re-oxidation of $NADPH_2$ may therefore be expected to result in greater utilization of the oxidative pentose-phosphate pathway. Important mechanisms for this re-oxidation include the biosynthetic reactions needing $NADPH_2$ as specific reducing agent.

8. Production of glycerol

Fermentative production of glycerol is uneconomic by comparison with the process based on the alkali saponification of fats.

Nevertheless, the microbiological method merits brief consideration, because of its interest as one of the earliest examples of a directed fermentation process, which was exploited in Germany for producing the glycerol needed for manufacture of explosives during the time of fat shortage in the First World War.

We have seen that continued glycolytic degradation of sugars demands some mechanism for continuous re-oxidation of the $NADH_2$ produced. Under aerobic conditions the re-oxidation is mainly effected through the respiratory chain of enzymes (see Chapter 12). When conditions are anaerobic, a variety of alternative hydrogen acceptors may be used; of these, the most important is the acetaldehyde produced by further metabolism of pyruvate, the end-product of glycolysis, during the course of alcoholic fermentation (see Chapter 10).

Glycerol production depends on modifying the normal course of alcoholic fermentation by removing acetaldehyde as soon as it is formed: this may be done either by addition of sodium bisulphite, which forms an addition compound with acetaldehyde, or by inducing a Cannizzaro reaction (one molecule of acetaldehyde is oxidized to acetic acid while another is reduced to alcohol) under alkaline conditions.

In either event, acetaldehyde is no longer available for re-oxidation of $NADH_2$, and its place as hydrogen acceptor is now taken by the dihydroxyacetone phosphate moiety of the triose phosphate produced during glycolysis (Fig. 9.2):

$$\begin{array}{c} CH_2O\,Ⓟ \\ | \\ CO \\ | \\ CH_2OH \end{array} + NADH_2 \rightleftharpoons \begin{array}{c} CH_2O\,Ⓟ \\ | \\ CHOH \\ | \\ CH_2OH \end{array} + NAD$$

The product is the L-isomer of glycerol-3-phosphate as opposed to the D-isomer that would have been produced if the D-glyceraldehyde-3-phosphate moiety of triose phosphate had been reduced instead.

8.1. *The soluble sulphite process*

To a medium containing sucrose 10–20% w/v; NH_4NO_3 0·5%; K_2HPO_4 0·075%; Na_2SO_3 4·0%; trace of $MgSO_4$, $7H_2O$; 1% w/v of filter-pressed yeast is added. The fermentation is conducted in large fermenters (up to 125,000 gal) and is allowed to proceed at 30°C for 48–60 hr. Gaseous evolution of carbon dioxide leads to bisulphite formation, according to the equation

$$Na_2SO_3 + CO_2 + H_2O \longrightarrow NaHSO_3 + NaHCO_3$$

The bisulphite fixes acetaldehyde [as the addition compound $CH_3CH(OH)SO_3Na$] and at the same time serves as an antiseptic that diminishes contamination of the broth.

The soluble sulphite process produces a broth with a high concentration of inorganic ions, but this may be avoided by substitution of ammonium sulphite if precautions are taken to control carefully the pH of the broth at 6·8 by automatic addition of sulphur dioxide.

The broth is ready for harvest when the sugar is depleted. The harvested liquor is clarified and distilled to remove alcohol, acetaldehyde and any gaseous ammonia before being further concentrated and cooled to remove residual sodium sulphite by crystallization. Fermentation broths at harvest usually contain no more than about 4% w/v glycerol, but the crude concentrated solutions before removal of sodium sulphite may contain up to 18% w/v glycerol.

8.2. *The alkaline process*

The procedure for fermentation is similar to that just described, but maintenance of an alkaline pH by addition of solid sodium carbonate (soda ash) at intervals during the first 24 hr (total amount added about 3·5 w/v) ensures removal of acetaldehyde in a Cannizzaro reaction and thus obviates the need to add soluble sulphite. The alkaline pH makes it necessary to use specially selected strains of yeast that are able to degrade carbohydrate under these adverse conditions.

Glycerol yields in broth equal to about 25% w/v of the fermentable sugar have been obtained, but the problem of efficient recovery has not been completely solved.

9. Dextrans

Bacterial dextrans are typical examples of extracellular polysaccharides produced by many micro-organisms. Dextrans have many potential uses in commerce (e.g. food preservation, lipstick manufacture), but they are mainly produced for use in medicine: dextrans with molecular weights of about 75,000 or greater are used as plasma substitutes, but dextrans with low molecular weight up to about 8000 are used, after conversion to dextran sulphate, as anti-coagulants.

9.1. *Biosynthesis of dextrans*

An extracellular enzyme (dextransucrase or sucrose transglucosylase) formed by the micrococcus, *Leuconostoc mesenteroides* (NRRL B-512) is able to synthesize dextrans from sucrose as specific substrate, according to the overall equation

$$\underset{\text{sucrose}}{nC_{12}H_{22}O_{11}} \longrightarrow \underset{\text{dextran}}{(C_6H_{10}O_5)_n} + \underset{\text{fructose}}{nC_6H_{12}O_6}$$

The detailed mechanism of this reaction has not been reported, but it seems probable that the process conforms to the known methods already described for biosynthesis of other natural polysaccharides from nucleoside diphosphate sugars as glycosyl donors.

$$\text{Nucleoside-DP-sugar} + \text{acceptor} \longrightarrow \text{nucleoside-DP} + \text{acceptor-glycoside}$$

cf. UDP-glucose + acceptor
⟶ UDP + acceptor-glucoside

Glucose or fructose in the free state cannot be used for dextran biosynthesis, and the utilization of the specific substrate, sucrose, does not involve phosphate in any way. Preliminary transformation of the substrate must therefore exclude the possibility of hydrolysis to glucose and fructose, and similarly phosphorolysis to glucose-1-phosphate and fructose must also be excluded. It is therefore possible that the mechanism in *L. mesenteroides* for preliminary transformation of sucrose may be similar to an enzymic reaction observed in wheat germ,

$$\text{UDP-glucose} + \text{fructose} \overset{Mg^{++}}{\rightleftharpoons} \text{sucrose} + \text{UDP}$$

The equilibrium of the wheat germ reaction is in favour of sucrose synthesis, but it must be presumed that the balance could be changed in favour of UDP-glucose if this product were depleted by instant utilization for dextran biosynthesis.

9.2. *Production of dextrans*

Clinical dextran can be produced either directly by batch fermentation or indirectly by use of the enzyme dextransucrase. Direct fermentation has the advantage of simplicity, but produces dextrans of various molecular weights so that the chain length must be reduced by hydrolysis until it conforms to specification (molecular weight of 70,000–80,000 in the U.S.A. and of 170,000–250,000 in the United Kingdom). The enzymic process, on the other hand, permits better control of the reaction and gives more uniform material with a mean molecular weight of 75,000.

Inoculum for either method of production is built up in several stages by serial transfers of about 5% v/v inoculum to successive batches of a good growth medium maintained at 25°C and terminating in transfer to a typical fermentation medium at the same temperature: sucrose 10% w/v; yeast extract 0·25%; K_2HPO_4 0·5%; $MgSO_4$, $7H_2O$ 0·02%; NaCl 0·1%; $(NH_4)_2SO_4$ 0·06%.

Agitation and aeration may be dispensed with in the earlier stages of inoculum production, but it is usual to provide mechanical agitation at least for the larger vessels, mainly as a means of temperature control.

Organic acids produced during the fermentation period cause the pH to fall, and the production of dextran increases the viscosity of the broth. The fermentation process is usually completed within 24 hr, by which time the pH may have fallen from an initial value of 7·0–7·5 to 4·5, with the viscosity around 400 to 700 centipoises. Some processes claim improved yields by buffering the medium to maintain pH 6–7 during the first 8 hr of the period of fermentation.

The enzymic process needs two main stages: aerobic production of the extracellular enzyme, dextransucrase, and enzymatic synthesis of dextran under reducing conditions.

Production of the enzyme needs only relatively low sugar concentration, a typical medium containing sucrose 2% w/v; cornsteep liquor 2%; KH_2PO_4 0·1%; mineral salts; initial pH 7·2. Optimum conditions for enzyme production are pH 6·7 and temperature 25°C, with mild aeration. Ammonium ions depress the yield of enzyme so that any necessary adjustment of pH is made with caustic alkali. Dextransucrase is formed at maximum rate at pH 6·7, but it is much stabler under more acidic conditions. Therefore it is usual practice to maintain the broth at pH 6·7 for the first 8 hr of the fermentation period and then to permit a gradual fall to pH 5·0 by the time of harvest at about 12–18 hr.

The enzyme is separated from the bacterial cells by centrifugation, the pH of the supernatant fluid is adjusted if necessary to 5·0, and the enzyme may then be stored at 15°C for as long as 30 days. Expected yield of enzyme is about 40 dextransucrase units per ml (1 unit of enzyme will convert in a mixture of 10% sucrose and M/20 acetate buffer at pH 5·0, 1 mg sucrose to dextran in 1 hr at 30°C).

Efficient use of dextransucrase is greatly modified by the nature of the glucosyl acceptor employed. Several different sugars, termed primers, will function as acceptors, but best results are obtained with about 2% w/v of material having a low molecular weight (materials that need more than 50% methanol for precipitation and having molecular weights of 15,000–20,000) remaining from fractionation of broth from previous batches.

As substrate, 10% w/v sucrose is supplied, and the incubation is conducted at 15°C or below in the presence of about 30 units/ml dextransucrase. Increasing the concentration of sucrose from 5% to 20% diminishes both the molecular weight of the product and its yield, so that the figure of 10% is a compromise. Similarly, both enzyme concentration and temperature affect the molecular weight of the product: in general, the molecular size of the product diminishes with increase in enzyme concentration, and the lower molecular weight dextrans also tend to predominate at 15°C, whereas the proportions of large and relatively small molecules are about equal at 30°C.

Dextrans produced by either the direct fermentation process or by the enzymic reaction are isolated from clarified liquors by precipitation with an equal volume of methanol, the organic acids and spent methanol being left behind in the supernatant.

10. The streptomycin fermentation

10.1. *Structure of streptomycin*

The antibiotics, streptomycin and mannosidostreptomycin, are glycosides of streptidine (1,3-diguanido-2,4,5,6-tetrahydroxycyclohexane), a cyclitol having the scyllo (all-*trans*) configuration (Fig. 9.6). The aglycone, streptidine, is in α-glycosidic linkage with an unusual methyl sugar, L-streptose, and this, in turn, forms an α-glycosidic bond with *N*-methyl-L-glucosamine. Finally, streptomycin may be joined by an α-glycosidic bond to D-mannose in mannosidostreptomycin.

The streptose residue may be modified by formation of primary alcohol groups to give related antibiotics: oxidation of the methyl group gives hydroxystreptomycin, and reduction of the aldehyde group forms dihydrostreptomycin.

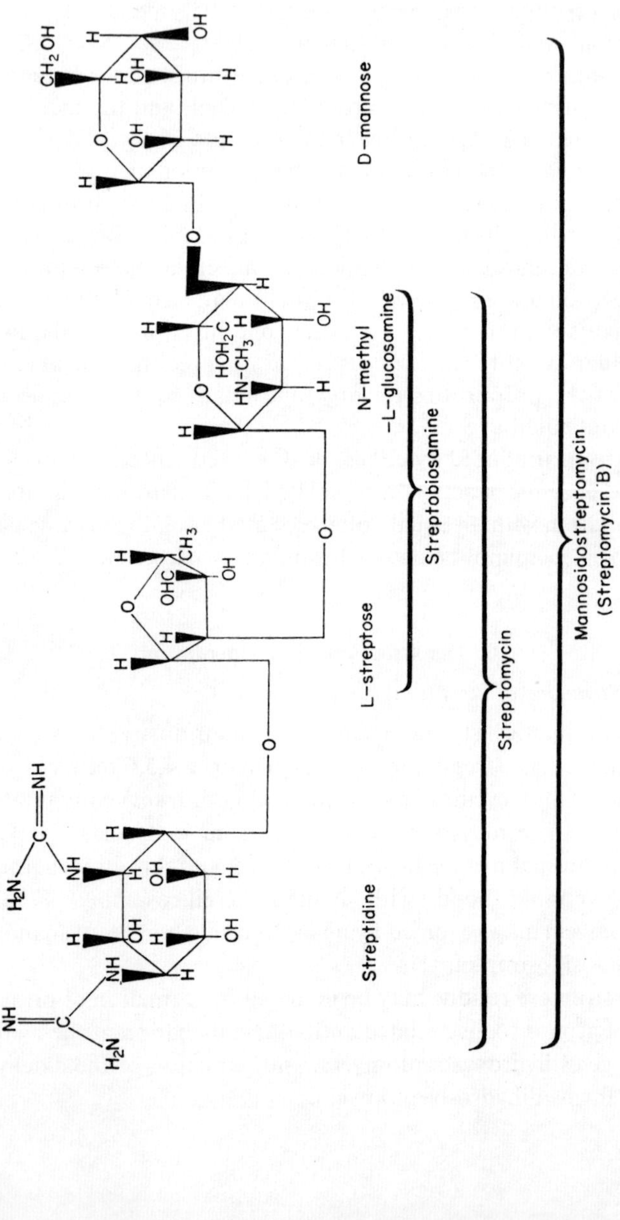

FIG. 9.6. Structure of streptomycin and mannosidostreptomycin.

10.2. *Biosynthesis of streptomycin*

The streptidine, L-streptose and *N*-methyl-L-glucosamine residues of streptomycin are equally derived from glucose; the sequence of formation of the α-glycosidic bonds has not been elucidated.

Arginine appears able to donate the two guanidine groups of the streptidine residue, a conclusion supported by the observed incorporation of radioactive carbon dioxide into the carbon atom of each guanidine side-chain, presumably by prior conversion to carbamyl phosphate, an intermediate in the biosynthesis of arginine.

The *N*-methyl group of *N*-methyl-L-glucosamine is derived by the usual process of *N*-methylation (see Chapter 14).

The streptidine residue appears to be derived directly from glucosamine-6-phosphate by ring closure, giving scyllo-inosamine with subsequent amination.

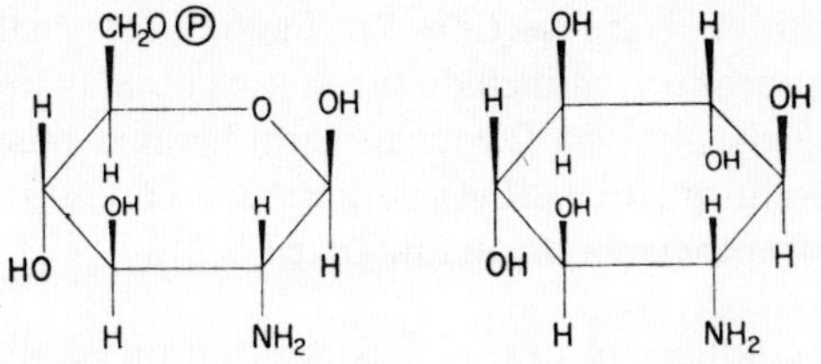

glucosamine – 6 – phosphate scyllo – inosamine

Nothing certain has been reported about the mode of biosynthesis of the streptose skeleton, but its ultimate source is glucose: the *C*-formyl group probably arises from C-3 of glucose.

It seems probable that epimerization of sugars may be involved in biosynthesis of streptomycin and these transformations would be expected to occur at the nucleoside diphosphate sugar level. It may be that similar intermediates are involved in formation of the α-glycosidic linkages.

10.3. *Production of streptomycin*

In view of the method of biosynthesis, it is not surprising that most commercial fermentation media for production of streptomycin by high-yielding strains of *Streptomyces griseus* provide glucose as the principal source of carbohydrate, although many less satisfactory sugars (fructose, galactose, lactose, maltose, xylose), sugar alcohols (D-mannitol) and polysaccharides, such as starch, are able to support good growth. It is important to note that streptomycin-producing strains of *S. griseus* are unable to use sucrose, the trisaccharide raffinose or the tetrasaccharide stachyose.

Nitrogen made continuously but slowly available from complex organic ingredients is to be preferred to ammonium salts or nitrates, largely because of the greater ease of pH control.

Carbohydrate breakdown can occur in *S. griseus* strains either by the Embden–Meyerhof–Parnas glycolytic pathway or by the Warburg–Dickens oxidative pentose-phosphate pathway; the extent to which either route is utilized will depend on the environmental conditions (see section 7). Favourable conditions for streptomycin production include the factors already described for limitation of glycolysis in favour of carbohydrate breakdown via the pentose-phosphate pathway: optimum conditions for production of streptomycin are summarized below.

(a) Adequate but not excessive concentration of glucose.
(b) Low concentration of inorganic phosphate (0·006% w/v to 0·02% soluble phosphate is optimum).
(c) Continuous but limited concentration of available nitrogen to prevent excessively rapid protein synthesis.
(d) Good aeration.
(e) Good control of pH within the range 7·0–8·0.
(f) Good temperature control within the range 27–29°C.

A typical fermentation medium for production of streptomycin may contain: glucose 2·5% w/v; extracted soyabean meal 4·0%; dried distillers' solubles 0·5%; sodium chloride 0·25%. An adequate amount of phosphate is provided by the distillers' solubles.

Adverse effects of excess soluble phosphate can sometimes be overcome by inclusion of calcium or magnesium salts in the medium: these salts presumably work by precipitation of insoluble phosphates. It is important to note that calcium and magnesium ions have only limited utility for this purpose, because they would interfere with the use of a cation exchange resin (see Chapter 2) for adsorption of streptomycin ions during the extraction process.

Three phases can be distinguished in the streptomycin fermentation. The first is characterized by release of ammonia from soya bean proteins, so that the pH rises steadily from an initial value of 6·7 to about 7·6 at the end of 20 hr. Concomitant growth with utilization of glucose and nitrogen will prevent excessive rise of pH during this period. The second is the productive phase characterized by a slow decrease in pH to about 6·7 accompanied by a rapid increase in streptomycin titre. Finally, in the third phase, the mycelium disintegrates or becomes lysed, with cessation of streptomycin production and rise of pH to between 8·0 and 8·5.

10.4. *Mannosidostreptomycin*

The α-mannoside of streptomycin, known as mannosidostreptomycin or streptomycin B, has a D-mannose residue linked by an α-glycosidic bond to the C-4 of *N*-methyl-L-glucosamine. Mannosidostreptomycin is produced in all streptomycin fermentations and may account for 25–40% of the total ferric-maltol reacting antibiotics in the early stages of fermentation. Mannosidostreptomycin has only about one-fifth of the antibacterial activity of streptomycin, so that the aim of the production process is to ensure that the amount of the α-mannoside does not exceed 10% of the total streptomycin at the time of harvest.

Mannosidostreptomycin tends to break down to streptomycin under the influence of an adaptive enzyme, mannosidase, during the later stages of fermentation. Mannosidase production can be induced (see Chapter 8) by adding either yeast mannan (a constituent of distillers' solubles) or 0·1% w/v α-methylmannoside. Unfortunately the action of the enzyme is also readily inhibited. Environmental conditions affecting the enzymic breakdown of mannosidostreptomycin to streptomycin may be summarized as follows:

(i) Heavy metals inhibit enzymic action.
(ii) Deficient aeration inhibits the action.
(iii) pH 8·0 is optimum for mannosidase activity, but this degree of alkalinity cannot be permitted until biosynthesis of streptomycin is almost completed.
(iv) Sugars have a marked effect on synthesis of the enzyme: more glucose than 0·5–1·0% in the medium completely inhibits formation of mannosidase.
(v) Increase of temperature up to about 32°C may accelerate the action of mannosidase, but this elevated temperature is permissible only after the biosynthesis of streptomycin has been completed.

Clearly, optimum conditions for mannosidase action can be fulfilled only near the termination of the fermentation period. The period of mannosidase synthesis is even more critical, because formation of the enzyme must be induced during the relatively short period that elapses between when the glucose concentration falls to below 0·5–1·0% and when the synthesis of protein (and therefore of enzyme) fails.

Suggested literature for further reading

1. Bixler, G. H., Hines, G. E., McGhee, R. M. and Shurter, R. A., Dextran, *Ind. Engng. Chem.* **45**, 692–705 (1953).
2. Candy, D. J., Blumson, N. L. and Baddiley, J., The biosynthesis of streptomycin: incorporation of ^{14}C-labelled compounds into streptose and *N*-Methyl-L-glucosamine, *Biochem. J.* **91**, 31–5 (1964).
3. Candy D. J. and Baddiley, J., The biosynthesis of streptomycin: the origin of the C–formyl group of streptose, *Biochem J.* **96**, 526–9 (1965).
4. Hehre, E. J., Studies on the enzymatic synthesis of dextran from sucrose, *J. biol. Chem.* **163**, 221–33 (1964).

5. HELLMAN, N. N., TSUCHIYA, H. M., ROGOVIN, S. P., LAMBERTS, B. L., TOBIN, R., GLASS, C. A., STRINGER, C. S., JACKSON, R. W. and SENTI, F. R., Controlled enzymatic synthesis of dextran: conditions for producing clinically suitable molecular weight, *Ind. Engng. Chem.* **47,** 1593–8 (1955).
6. HESTRIN, S. and AMINERI-SHAPIRO, S., The mechanism of polysaccharide production from sucrose, *Biochem. J.* **38,** 2–10 (1944).
7. HOCKENHULL, D. J. D., FANTES, K. H., HERBERT, M. and WHITEHEAD, B. K., Glucose utilization by *Streptomyces griseus*, *J. gen. Microbiol.* **10,** (3) 353–69 (1954).
8. HOCKENHULL, D. J. D., ASHTON, G. C., FANTES, K. H. and WHITEHEAD, B. K., Actinomycete metabolism: α-phenyl-mannosidase of *Streptomyces griseus*, *Biochem. J.* **57** (1) 93–8 (1954).
9. HOLLMAN, S. (translated and revised by Touster, O.), *Non-glycolytic Pathways of Metabolism of Glucose*, Academic Press, New York and London, 1964.
10. HORNER, W. H., Biosynthesis of streptomycin. II. Myo-inositol, a precursor of the streptidine moiety, *J. biol. Chem.* **239,** 2256–8 (1964).
11. LELOIR, L. F., Nucleoside diphosphate sugars and saccharide synthesis, *Biochem. J.* **91,** 1–8 (1964).
12. LOCKWOOD, A. R. and SWIFT, G., U.S.A. patent 2565507; 1951.
13. NEELY, W. B., Dextran: structure and synthesis, *Adv. Carbohyd. Chem.* **15,** 341–69 (1960).
14. NEUFELD, E. F. and HASSID, W. Z., Biosynthesis of saccharides from glycopyranosyl esters of nucleotides (sugar nucleotides), *Adv. Carbohyd. Chem.* **18,** 309–56 (1963).
15. STACEY, M., Enzymic production of bacterial polysaccharides, *Nature, Lond.* **149,** 639 (1942).
16. WALKER, M. S. and WALKER, J. B., Evidence for participation of a phosphorylated derivative of streptidine in streptomycin biosynthesis, *Biochem. Biophys. Acta.* **97,** 397–8 (1965).
17. WILKIN, G. D. and RHODES, A., Observations on the morphology of *Streptomyces griseus* in submerged culture, *J. gen. Microbiol.* **12** (2) 259–64 (1955).

CHAPTER 10

Fermentation Products Formed from Pyruvate by Facultative Anaerobes

GLYCOLYSIS of sugars to pyruvate has been described in the preceding chapter. Pyruvic acid is a key compound in the metabolism of micro-organisms but the nature of its subsequent transformation will depend on the tissue and on the environmental conditions. Complete oxidation to carbon dioxide and water via the citrate cycle (see Chapter 12) is a common fate, much of the energy of respiration being stored in the form of energy-rich phosphate bonds until required for biosynthetic reactions. This pathway requires a preliminary but irreversible oxidative decarboxylation of pyruvate to give carbon dioxide and acetyl-coenzyme A or active acetate, but it can operate normally only under aerobic conditions. When oxygen is temporarily deficient or under anaerobic conditions, alternative pathways for metabolism of pyruvate may be utilized. Processes employed by facultative anaerobes are described below; the anaerobic routes used by butyric acid bacteria will be described in the next chapter.

Knowledge of the mechanism for oxidative decarboxylation is essential to an understanding of the means employed by facultative anaerobes for metabolism of pyruvate by alternate routes when oxygen is deficient.

1. Oxidative decarboxylation of pyruvic acid

The mechanism summarized in Fig. 10.1 is generally applicable to the oxidative decarboxylation of α-keto acids. A carboxylase

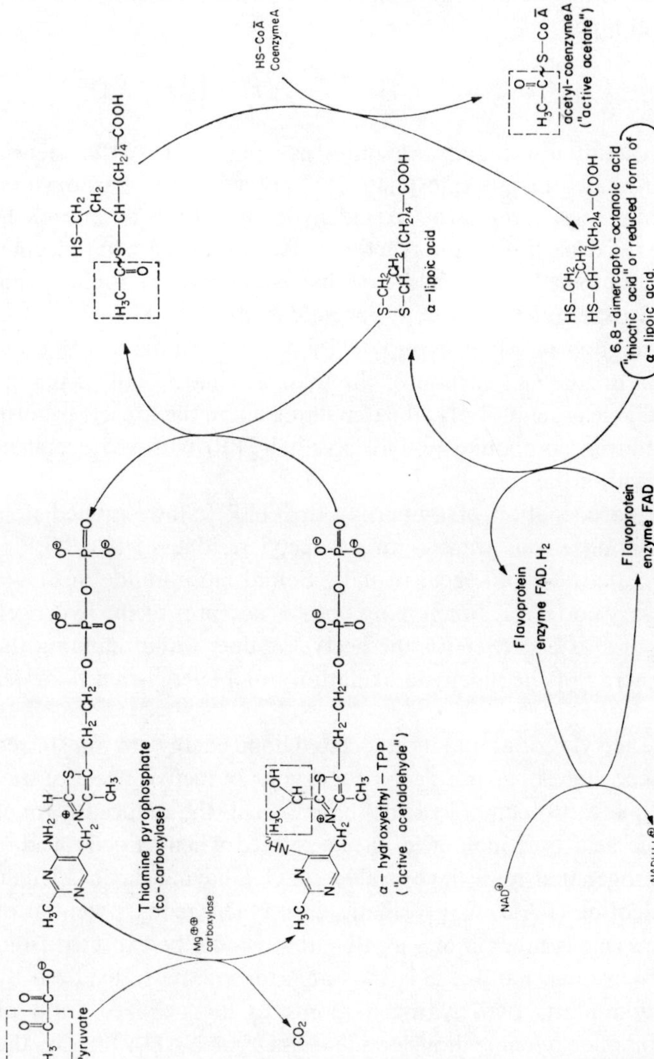

FIG. 10.1. Oxidative decarboxylation of pyruvate.

enzyme catalyses a preliminary decarboxylation of the following general type:

$$CH_3.CO.COO^{\ominus} + H^{\oplus} \longrightarrow CH_3.CHO + CO_2$$

The reaction also requires thiamine pyrophosphate (TPP, aneurin diphosphate, the pyrophosphate of vitamin B_1, or co-carboxylase) and $Mg^{\oplus\oplus}$. A complex of acetaldehyde and TPP is an intermediate in the decarboxylation reaction. This intermediate has recently been identified as α-hydroxyethyl-2-thiamine pyrophosphate (α-hydroxyethyl-TPP or active acetaldehyde).

Formation of α-hydroxyethyl-TPP depends on the fact that the proton on the C-2 carbon of the thiazole ring of TPP is readily dissociable at neutral pH: the remaining anion then reacts to form an addition compound (α-hydroxyethyl-TPP) with the acetaldehyde within the complex.

Dehydrogenation of α-hydroxyethyl-TPP follows immediately with simultaneous transfer of an acetyl residue so that TPP is regenerated. α-Lipoic acid (usually bound in an amide link) is a necessary co-factor, functioning both as acceptor of the hydrogen atom and as a carrier for the acetyl residue, while retaining the free energy of the aldehyde oxidation in an energy-rich thioester bond.

The acetyl residue and its associated bond energy are transferred to coenzyme A, giving acetyl-coenzyme A (active acetate) and 6:8-dimercapto-octanoic acid (thioctic acid, the reduced form of α-lipoic acid). α-Lipoic acid is regenerated when thioctic acid is dehydrogenated by a flavoprotein enzyme having flavin adenine dinucleotide (FAD) as prosthetic group. The redox potential of this enzyme is much more negative than would be expected from its flavoprotein nature; it is thus able to become re-oxidized by donation of its two hydrogen atoms to the oxidized form of nicotinamide adenine dinucleotide ($NAD^{\oplus}$ or NAD). Finally, the reduced form of nicotinamide adenine dinucleotide ($NADH + H^{\oplus}$ or $NADH_2$) is re-oxidized by the normal respiratory chain of enzymes to permit continuation of the cycle of oxidative decarboxylation. The overall reaction, summarized in Fig. 10.1, may be represented by the equation:

$$CH_3.CO.COOH + NAD^{\oplus} + HS.Co\overline{A} \longrightarrow CH_3.CO{\sim}S.Co\overline{A} + CO_2 + NADH + H^{\oplus}$$

It can be seen that continued operation of the system depends upon the enzymes of the respiratory chain being able to re-oxidize the mole of $NADH_2$ formed in each cycle. These respiratory enzymes normally function under aerobic conditions, permitting ultimate transfer of the equivalent of two hydrogen atoms to atmospheric oxygen (see Chapter 12).

2. Metabolism of pyruvate under conditions of partial or complete anaerobiosis

In many micro-organisms oxidative decarboxylation of pyruvate can proceed even under conditions of temporary anaerobiosis. In these circumstances, pyruvic acid itself, in the presence of lactic dehydrogenase, can act as the hydrogen acceptor in re-oxidation of $NADH_2$, with the result that one molecule of pyruvic acid is oxidatively decarboxylated to acetyl-coenzyme A and another is reduced to lactic acid:

$$2CH_3.CO.COOH + \tfrac{1}{2}O_2 \longrightarrow CH_3.CHOH.COOH + CH_3CO{\sim}S.Co\overline{A} + CO_2$$

Formation of lactic acid from pyruvic acid is a reversible reaction, but the equilibrium will be well over towards lactic acid under conditions of rapid glycolysis, so that this method for re-oxidation of $NADH_2$ is no longer adequate and oxidative decarboxylation of pyruvic acid must fail.

In some micro-organisms, and particularly in yeasts, the enzyme carboxylase is abundant: this facilitates dissociation of the intermediate α-hydroxyethyl-TPP without transfer of the H atom to α-lipoic acid as in oxidative decarboxylation. Acetaldehyde is released, and decarboxylation of pyruvic acid can continue under anaerobic conditions. The overall equation for this decarboxylation is given by:

$$\underset{\text{pyruvic acid}}{CH_3.CO.\boxed{COO}\,H} \longrightarrow \underset{\text{acetaldehyde}}{CH_3.CHO} + \boxed{CO_2}$$

Magnesium ions are required for this process as well as for oxidative decarboxylation, because this metal is a co-factor for the TPP (co-carboxylase) common to both systems.

The acetaldehyde formed by decarboxylation of pyruvic acid and dissociation of the active acetaldehyde complex is detoxified by reduction, especially in yeasts, to ethanol by the agency of an NAD-specific, sulphydryl-containing enzyme designated alcohol dehydrogenase, working in reverse. The 2H required for reducing each mole of acetaldehyde are obtained by re-oxidation of the $NADH_2$ generated during the glycolytic breakdown of carbohydrate (see Chapter 9) at the time when hydrated glyceraldehyde-3-phosphate is dehydrogenated to glyceric-acid-1:3-diphosphate by triosephosphate dehydrogenase and the co-factors NAD and inorganic phosphate. Formation of ethanol thus serves a dual purpose: the process removes acetaldehyde, the potentially toxic product of pyruvic acid decarboxylation, and at the same time permits continued glycolysis under the anaerobic conditions that might otherwise cause failure by accumulation of all available nicotinamide adenine dinucleotide in the reduced form.

The active acetaldehyde complex (α-hydroxyethyl-TPP) formed during the thiamine pyrophosphate-dependent decarboxylation of pyruvate thus functions as a key metabolite (Fig. 10.2). Normal oxidative decarboxylation yielding acetyl-coenzyme A (route *A*) occurs under aerobic conditions, but may be replaced under conditions of temporary anaerobiosis by dismutation of 2 moles of pyruvic acid to 1 mole of acetyl-coenzyme A and 1 mole of lactic acid (route *B*). In some micro-organisms the active acetaldehyde complex can dissociate to yield free acetaldehyde (route *C*),

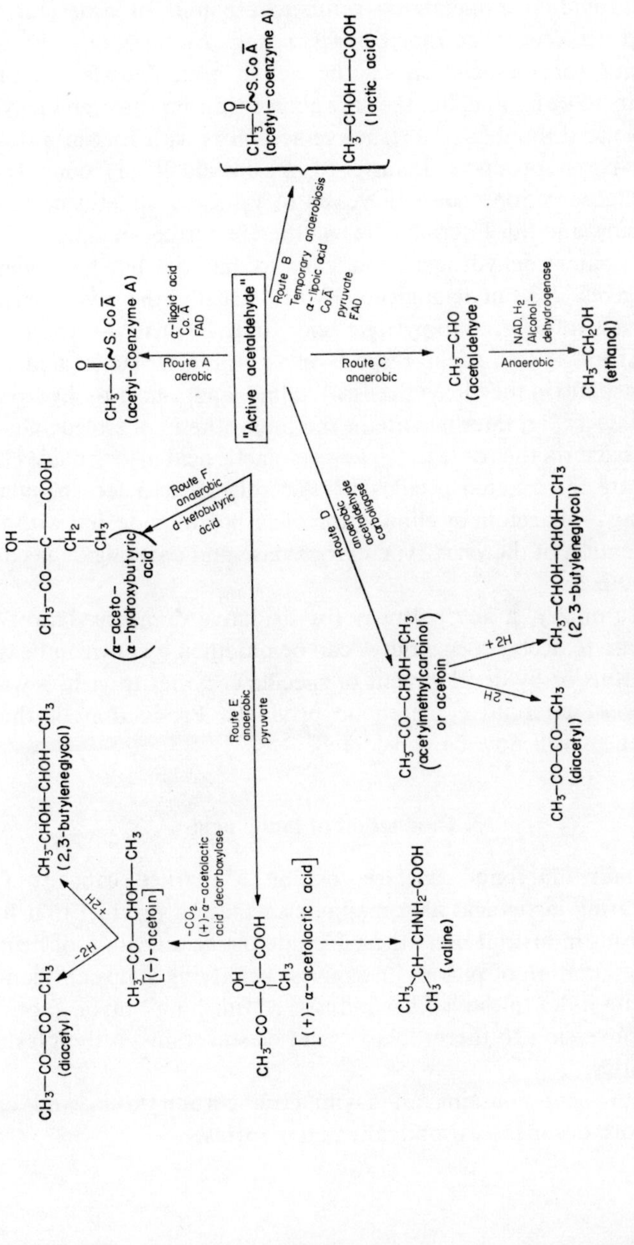

FIG. 10.2. Alternative routes for metabolism of active acetaldehyde (α-hydroxyethyl-2-thiamine pyrophosphate).

thus providing a metabolic route to ethanol. In other micro-organisms (e.g. *Neurospora crassa* and *Escherichia coli*) the capacity for dissociation of the active acetaldehyde complex appears to be limited, but they are able to transfer the acetaldehyde residue to a number of alternative acceptors with formation of a new range of products. Transfer to free acetaldehyde (route *D*) by carboligase action occurs in yeasts to yield acetylmethylcarbinol (acetoin), and this intermediate can then be further metabolized in some strains: dehydrogenation leads to diacetyl, but hydrogenation gives 2:3-butylene-glycol. More usually the acetaldehyde residue is transferred to pyruvic acid (route *E*) or to α-ketobutyric acid (route *F*) to yield respectively (+)-α-acetolactic acid, an intermediate in the biosynthesis of valine, and α-aceto-α-hydroxy-butyric acid, an intermediate in the biosynthesis of isoleucine. In some bacteria the route to (+)-α-acetolactic acid also provides the substrate for a stereospecific (+)-α-acetolactic acid decarboxylase forming (—)-acetoin by elimination of carbon dioxide, but without intervention of the yeast-type carboxylase and carboligase needed for route *D*.

In summary, a key pathway for oxidative decarboxylation of pyruvate to acetyl-coenzyme A can be modified by environmental conditions or by development of specific enzymes to yield a wide variety of alternative metabolic products. Production of these substances will now be considered.

3. Production of lactic acid

Certain fungi and bacteria exhibit a marked capacity for elaborating lactic acid as a major fermentation product that has numerous industrial uses: these include the acidification of brines for preservation of pickles and olives; acidifying soups and jams; deliming hides in the leather industry; "finishing" rayon fabrics; and conversion to thermoplastic and elastic resins in the plastics industries.

Lactic acid contains an asymmetric carbon atom and can therefore occur in two optically active forms:

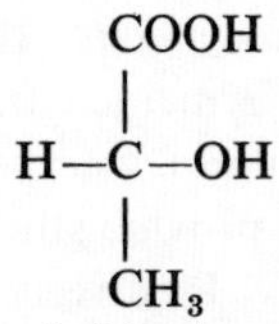

D-(—)-lactic acid

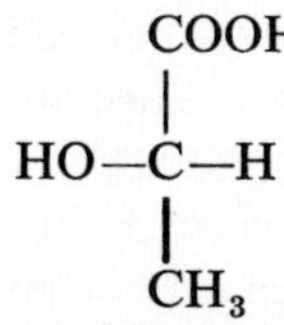

L-(+)-lactic acid

The particular isomer formed from pyruvic acid depends upon the nature of the dehydrogenase found in the micro-organism used: homofermentative bacteria (producing lactic acid as the principal end product), in common with the mould *Rhizopus oryzae*, contain an L-lactic acid dehydrogenase and produce L(+)-lactic acid, but heterofermentative bacteria (producing volatile acids and carbon dioxide in addition to lactic acid as principal end products) contain a D-lactic acid dehydrogenase and therefore form D(—)-lactic acid. Although L(+)-lactic acid is the end product of the *R. oryzae* fermentation, it is usual for bacterial fermentations to yield mainly the optically inactive racemate. Racemization of the first formed and specifically D(—)- or L(+)-lactic acids is brought about by the activity of an enzyme, racemase, produced by several bacteria, including some lactic acid bacteria. Racemase activity depends on nicotinic acid, and its action in the fermentation can therefore be controlled to some extent: complete control is not possible because nicotinic acid is also an essential requirement for the growth of DL-lactic acid bacteria.

Industrial fermentations for lactic acid are relatively expensive because of the difficulties encountered in recovering and purifying the end product.

The lactic acid bacteria (Lactobacillaceae) are gram-positive bacteria in the genera *Streptococcus*, (*Pediococcus*), *Leuconostoc* and *Lactobacillus*. They are anaerobic or micro-aerophilic organisms with rather exacting nutritive requirements: many require members of the vitamin B complex and certain amino acids as supplements for growth. Industrial fermentations depend mainly on the use of homofermentative lactic acid bacteria, including *Lactobacillus delbrueckii*, *L. bulgaricus*, *L. pentosus*,

L. casei, *L. leichmanii* and *Streptococcus lactis*. The organism is selected to suit the carbohydrate to be fermented and the temperature of fermentation. Carbohydrate sources, whether used directly or after preliminary treatment, usually provide glucose, sucrose or lactose in concentration of 5–20% w/v. Relatively high temperatures are employed: *Lactobacillus delbrueckii* and *L. bulgaricus* are used at about 45–50°C and *Streptococcus lactis*, *Lactobacillus pentosus* and *L. casei* at about 30°C.

The fermentation vats need no aeration, owing to the micro-aerophilic nature of the micro-organisms, but the corrosive nature of the product precludes the use of vessels made from iron or copper alloys.

The fermentation is begun at about pH 5·8–6·0; the pH is continuously controlled by any of the usual methods at about this level to counteract the acidity of the lactic acid. The fermentation is complete when the sugar is almost exhausted: this is usually at the end of 1–6 days, with yields of about 90% from the sugar used. Contamination is not a serious problem, as it tends to be prevented by the deficiency of oxygen and by the high temperature.

4. Alcoholic fermentations

4.1. *Industrial alcohol*

Ethyl alcohol can be produced by fermentation of any carbohydrate material containing a fermentable sugar or a polysaccharide that can be hydrolysed to a fermentable sugar. Yeasts are used almost exclusively for these fermentations, but the species employed is dictated by the nature of the carbohydrate raw material: *Saccharomyces cerevisiae* strains are used when hexoses are to be fermented, but *Candida pseudotropicalis* or *C. utilis* is needed for fermentation of lactose or of pentoses respectively. Yields of ethyl alcohol are frequently greater than 90% of the sugar consumed by high-yielding strains having a high tolerance for alcohol. The fermentation time can be varied within wide limits, depending on temperature, sugar concentration and strain.

The production of ethyl alcohol is basically similar in the inoculum-development stages to the production of fodder yeast (see Chapter 8), but the end-product is dependent upon the operation of a metabolic shunt mechanism, and therefore the normal process of oxidative decarboxylation of pyruvate, occurring when carbohydrate is katabolized aerobically under optimum growth conditions, must be diverted to acetaldehyde, so as to terminate in ethanol (Fig. 10.2: route *C*). Diminution of the normal route by deficient aeration with simultaneously increased glycolysis is achieved by means of high initial sugar concentration (10–18% w/v) and plentiful phosphate.

Production of industrial alcohol was, until 1930, entirely by the fermentation process. The main raw material was either blackstrap molasses or, less frequently, high-test or beet molasses, but increased costs of these raw materials and the rapid development of synthetic ethanol production has now rendered the fermentation process uneconomic, except for certain specialized purposes. Industrial alcohol will probably continue to be obtained on a diminished scale as the end-product of processes designed to reduce the biological oxygen demand (B.O.D.) of some industrial wastes, including whey and the sulphite waste of paper pulp mills (see Chapter 16). The large amount of carbon dioxide evolved from decarboxylation of pyruvate during the fermentation period is sometimes recovered and converted to solid carbon dioxide or other by-products. Industrial alcohol produced from maize is reputed to have the advantage of freedom from foreign odours and flavours and may therefore continue to find some specialized outlets in the pharmaceutical and allied industries.

4.2. *Alcoholic beverages*

These products of alcoholic fermentations originated in spontaneous fermentation processes of great antiquity, but it is only in recent years that modern methods of industrial microbiology have been applied to their manufacture. Production of alcoholic beverages, unlike industrial alcohol fermentation, does not suffer competition from synthetic products, because the "character" of the beverage is dependent upon the interactions between a variety of biological factors that have not yet been defined in chemical or physical terms.

The type of beverage produced is determined by the nature of the plant material employed for fermentation: beer types are obtained from cereal grains, wines from a variety of different fruits, though the best from grapes, and mead-types from the fermentation of honey or the sap of suitable plants. In all these processes the method of preparing the fermentation medium is a factor of prime importance.

4.2a. *Beers.* The process of brewing or beer-production can be considered in four stages: preparation of the fermentation medium or wort, fermentation, maturation and storage.

Cleaned barley grain is steeped for up to 2 days in water, drained and permitted to germinate on the malting-floor or in revolving drums. Growth of the sprouted grain is stopped after about 5 days by careful drying in a kiln: the resultant malt contains amylolytic and proteolytic enzymes and is stored until required for use. Recent work has shown that the plant growth regulator, gibberellic acid, or preferably the potassium salt, included in the steep-water at a concentration of 0·25–5·0 mg/l., reduces the time needed for germination; it also diminishes the malting loss, possibly by stimulation of enzyme activity.

Malt provides starch and protein as the basis of the fermentation medium, and these substances are converted to fermentable sugars and dextrins and to amino acids and other products of protein degradation by the action of indigenous amylolytic and proteolytic enzymes, respectively, when the malt is incubated in warm water during the process of mashing. The temperature and quality of the water employed in mashing plays a large part in determining the nature of the brew. Proteolytic enzymes of malt exhibit optimum activity at about 40°C, but the α- and β-amylases that convert amylopectin (a branched-chain glucose polymer) and amylose (a straight-chain glucose polymer), the constituent polysaccharides of starch, to dextrins and maltose, respectively, work best at higher temperatures (about 57–65°C for β-amylase, a little higher for α-amylase). Temperature control during the mashing process can therefore determine the relative proportions of non-fermentable dextrins to fermentable sugars and also the extent of proteolysis. Worts having a high proportion of dextrins that cannot be fermented by yeasts afford beers of low alcohol content, but those containing large amounts of sugars give high alcohol beers known as ales. The extent of proteolysis mainly affects flavour, physical properties and shelf-life of the beer: high molecular weight nitrogenous compounds predominate at above-optimum conversion temperatures (about 60°C) and provide good flavours, foam and foam stability.

Countries, such as the U.S.A., that utilize malt prepared from barley of relatively high nitrogen content employ supplementary carbohydrates as substrates for the excess amylases: materials known as malt adjuncts (non-germinated grains, sugars, syrups and, less frequently, soya beans, cassava or potatoes) afford beers of pale colour and lower protein content, which are relatively less satiating and more stable.

The mineral content of the water used in mashing is important: soft waters that contain 250–500 μg/ml of calcium sulphate are suitable for light Pilsen-type beers and for the English Burton ales, but water of high carbonate content is responsible for the darker colour and stronger flavour of stouts and Munich beers, because higher pH favours extraction of acidic colouring matters, tannins and bitter resins from the husks of the barley grains.

The mashing process is completed by filtration, so as to separate the clear wort from the insoluble grain residues, which, after further repeated extraction or "sparging" with water at about 75°C, contain less than 1% of soluble extract. These spent residues are made available as cattle fodder, and the extract is used for supplementing the clarified wort.

The wort is boiled for about 2 hr with addition of hops (the dried female flowers of the hop plant) at intervals. Boiling serves to sterilize and concentrate the wort, to inactivate its constituent enzymes, partially to caramelize the sugar and to extract from the hops the hop resins and hop oils, which impart the characteristic aroma and bitter flavour to beer, together with some antibacterial substances, which help to prevent infection. Much of the wort protein is precipitated by tannins extracted from the malt and from the boiled hops: the insoluble sludge is then removed with the spent hops. The wort is cooled below 22°C as an aid to further precipitation of protein–tannin complexes, in an effort to diminish formation of the protein-haze that results when precipitation of protein continues in the finished beer. Finally, the clarified wort is transferred aseptically to the fermentation vessels.

Beer-type fermentations are mainly induced by strains of *Saccharomyces cerevisiae* and *S. carlsbergensis*. Top-fermenting and bottom-fermenting groups are recognized, depending respectively upon their characteristic behaviour in either rising to the surface or settling on the bottom of the fermentation tank during the period of active fermentation. The characteristic behaviours of top- and bottom-fermenting yeasts determine the type of fermenter that can be employed: bottom-fermenting yeasts may be fermented in closed fermenters, but open vats are necessary for the top-fermenting yeasts that are extensively used for ale production in the United Kingdom.

We have seen that the "character" of a beer is dependent to a large extent upon the methods and materials used for preparing the wort, but the strain of fermenting yeast is also of prime importance. Desirable properties in a brewery yeast are ability to produce a vigorous fermentation in wort; a degree of flocculence such that the yeast cells aggregate, thereby permitting cleansing of the brew at the desired attenuation when the correct proportion of sugars have been converted to alcohol; and ability to produce the fusel oil by-products (volatile acids and higher alcohols) that play a large part in imparting flavour and aroma.

Microbial spoilage can be troublesome in some breweries, because completely effective methods of sterilization need to be omitted because of their deleterious effect on the taste of the beverage. The problem is aggravated when the initial yeast inoculum (the pitching yeast) is a mixed culture of top yeasts obtained by skimming the yeast head from the vats containing previous brews. Beer is not very suitable as a medium for general bacterial growth, because of its low oxygen, sugar and amino acid contents, its relatively low pH of about 4 to 5 and the presence of the mildly antibacterial ethanol, fusel oil and hop antiseptics. Bacterial spoilage is therefore mainly caused by the micro-aerophilic lactic acid bacteria and the acetic acid bacteria that are able to metabolize ethanol: they cause spoilage due to turbidity and off-flavours, and they may cause serious fining difficulties when the finished beer is clarified by the coagulating action of isinglass or other colloidal fining agents. Wild yeasts, particularly *Rhodotorulae*, present the main problem, and the pitching yeast itself is probably the main source of contamination in vat fermentations with top yeasts. The problem is likely to diminish when the techniques of pure culture inoculum described in Chapter 4 are more extensively applied.

Three fermentation techniques are employed, in open vats, in closed batch fermenters or in continuous fermenters: the course of fermentation is basically similar in all three. Vats containing cooled wort are inoculated or "pitched" with the top yeast inoculum (2 g or more of moist yeast per litre of wort). After an initial lag period, growth of the yeast is rapid, and the fermentation quickly becomes anaerobic. Artificial cooling is applied to limit the rising temperature of the fermenting wort to 20°C. Yeast floats near the surface of the vats as the fermentation progresses, and this is skimmed off at intervals, the material taken from the most vigorous period of fermentation being used as inoculum for pitching subsequent brews. The fermentation is completed within 5–6 days. The beer, which still contains a small quantity of fermentable sugar, is cooled and removed to cool storage tanks or barrels, where it may be retained for up to several weeks, so as to permit slow secondary fermentation. During a further storage period, isinglass or other finings material is added to precipitate protein–tannin complexes and the residual yeast cells before the clarified beer is distributed in casks or, after pasteurization, in bottles.

Closed batch fermentations, such as are used for production of lagers by bottom yeasts, proceed in a similar manner but at lower temperature (5–10°C initially and up to about 15°C as the fermentation progresses) and over a longer period (7–12 days), then with prolonged storage at 0–3°C to permit the slow secondary fermentation known as lagering. The clarified beer is saturated under pressure with carbon dioxide (obtained from the gas evolved during fermentation) before filling out in the usual manner.

Continuous beer fermentations are a relatively new development that offer considerable potential for development. The principle differs little from the method of serial addition of nutrients, that has already been considered in Chapter 8 for production of commercial bakers' yeast, except that the rate of growth will now be slower because of the anaerobic environment. Problems involved in preparing the fermentation medium and preserving the purity of the yeast culture present greater difficulties, but these are likely soon to be overcome.

The method of dispensing draught beer is also undergoing change. In 1955 "container" beer was introduced in the United Kingdom and it has now gained widespread popularity because of its clarity and uniformly good condition. Most container beer is brewed in the conventional manner of the draught product but it is then matured in glass-lined tanks instead of wooden casks. Container beer is carbonated to a slightly higher degree than draught beer before being clarified, stabilized, and filled into aluminium alloy or stainless steel casks.

4.2b. *Wines.* Grape wines are the most important products formed by alcoholic fermentation of fruit or berries. Characteristics of the fermentation medium and the nature of the yeasts employed play vital roles similar to those already described for brewing.

Selected entire ripe, but not over-mature, grapes are crushed, and the liberated juice or "must" is transferred to the vats or to the closed fermenters. It is necessary to eliminate the stalks that support the berries because these contain tannins that would give an undesirable flavour to wine. The skins of the grapes contain much of the largely unidentified aromatic and flavouring substances, together with the anthocyanins (mainly oenin) that give the colour to black grapes. It is usual to include the skins of black grapes in the must used for producing red wine although extracts of skin are utilized as an alternative in some processes.

The juice has a pH of about 3·0 to 3·6, determined mainly by its organic acid content (tartaric and malic acids, together with a little citric acid), and contains 15–25% w/v of fermentable sugars (mainly glucose and fructose) together with small amounts of pentoses, pentosans, pectins and inositol. Adequate available nitrogen (mainly in the form of amino acids, with some peptides and purines) is present together with a plentiful supply of accessory food factors and mineral salts. The must is therefore a good growth medium for acid-tolerant micro-organisms and does not need mashing (as in brewing) to render fermentable the carbohydrates and nitrogenous compounds.

Absence of mashing and boiling processes, necessary in medium preparation for brewing, simplifies preparing the fermentation medium for wines, but presents additional hazards of infection. This is aggravated by the presence in the "bloom" of a miscellaneous collection of micro-organisms, comprising mainly yeasts, moulds and bacteria, in addition to the wine yeasts that often provide the natural inoculum characteristic of particular areas. The problem is solved by partial sterilization of the must with sulphur dioxide, bisulphites or metabisulphites: wine yeasts are somewhat resistant to these substances; consequently, if the concentration of inhibitor is carefully controlled, the population of micro-organisms remaining in the must is composed largely of wine yeasts. Much wine is produced by this type of spontaneous fermentation, induced by mixed inocula of naturally occurring yeasts sometimes acting in association with acid-tolerant-bacteria: in favourable years it can lead to production of particularly good wine.

Lack of rigid control of the inoculum has led to selection and culture *in vitro* of desirable strains of wine yeasts (mainly strains of *Saccharomyces cerevisiae*, var. *ellipsoideus*). Inoculum is prepared by propagation of the selected strain in sterile must under optimum growth conditions, with serial transfers at the logarithmic phase of growth until sufficient cells are available to permit a 1–3% v/v but never more than 5% inoculation of the bulk partially sterilized must. Pure culture inoculation offers the additional advantage that the must can be clarified by centrifugation or filtration before the fermentation is begun, because removal of the natural flora is not detrimental to this process.

The course of the fermentation is similar to that of other alcoholic fermentations: an initial period of rapid cell multiplication under aerobic conditions leads to a prolonged anaerobic fermentation, from which the final alcohol content is governed by the alcohol-tolerance of the yeast employed and by the initial sugar concentration of the must. Most table wines are limited to about 10% maximum alcohol content so that most of the must sugar is fermented before the concentration of alcohol inhibits fermentation.

It is necessary to dissipate the heat produced during the fermentation process in order to maintain optimum conditions for wine yeast fermentation and render the environment less suitable for contaminating lactic and acetic acid bacteria. Temperatures below 30°C are usually favoured, and it is found that aroma develops best below about 24°C. Fermentations are largely completed in 3–5 days, when a layer of grape tissues known as "pomace" will be present on the surface of the fermented liquor and sufficient tannins and pigments have been extracted from the skins by the ethanol produced during the fermentation.

The partly fermented wine is decanted from the pomace as free-run juice into storage tanks at about 21–29°C. A slow fermentation proceeds during the next 7–10 days, until the residual sugar has been consumed: meanwhile, the carbon dioxide produced is liberated from the tanks. When the fermentation is completed the new wine is decanted or racked from the sediment (lees) of yeast and particles of grape tissues to facilitate clarification and to avoid extraction of undesirable substances that might have an adverse effect on flavour.

The clarified wine is aged in completely filled and sealed tanks: these contain little oxygen, and the chance of contamination by acetic acid bacteria, the main spoilage micro-organisms of wine, is therefore diminished. The wine is racked at intervals, and the contents of the tanks are adjusted for evaporation loss by addition of wine as necessary. This process of repeated racking and filling aids clarification, loss of carbon dioxide and the absorption of some oxygen: tannins become oxidized, and aroma is developed by other oxidative processes that lead to ester formation. Wines may take from 2 to 5 years to age, depending upon type: some will have cleared naturally by the end of this period, but others may still need to be clarified by addition of fining materials (tannins, casein, gelatine, bentonite and the albumen from the swim bladder of sturgeon, known as isinglass), with subsequent filtration with filter aids before filling into containers for sale. Some wines are pasteurized in the bottle before distribution.

This basic process for production of table wines is modified in a number of ways after completion of fermentation to produce wines of different types.

Sparkling wines, such as champagne, are made by carbonation of wine under pressure. Traditionally, this is achieved by a secondary alcoholic fermentation induced in the bottled wine when sugar and a fresh inoculum of a champagne strain of *S. cerevisiae*, var. *ellipsoideus*, are added. The bottles are held under pressure, so that the carbon dioxide liberated during the secondary fermentation is retained to carbonate the wine and confer its sparkling character. Cheaper forms of sparkling wine are produced by conducting the secondary fermentation in closed fermenters. Sometimes carbonation with carbon dioxide under pressure is substituted for the secondary fermentation, but wines produced by this method soon lose their sparkle when the bottle is opened.

Fortified wines, consumed as aperitifs or as dessert wines, include sherry, port, madeira and vermouth: their alcohol content is usually increased to a maximum of about 20% by addition of brandy. Further variety is obtained by sweetening with geropiga (must that has been prevented from fermenting by addition of spirit) to give port, and by addition of mixed herb infusions, to give sweet or dry vermouths. The production of sherry requires a prolonged but slow secondary fermentation with a surface growth of "flor" yeast and then colouring and sweetening according to type.

4.2c. *Spirits*. These are produced by distilling various fermented liquors: whisky, brandy, gin, rum, vodka and numerous liqueurs are typical products. The fermentation processes are similar to other alcoholic fermentations, the characteristics of the spirit being largely determined by the raw material used to produce the fermented liquor, by the conditions of distillation and ageing and by the nature of the additives. Distillers invariably use strains of *S. cerevisiae* that are usually capable of greater attenuation of the fermentation and that have frequently greater tolerance towards ethanol. Some strains have a particular facility for formation of fusel oil (a mixture of higher alcohols formed by deamination and decarboxylation of amino acids), which contributes to the characteristic aroma and flavour of the product.

Brandy and rum are obtained from distillation of grape wine and fermented sugar juices, respectively. Gin is prepared by adding juniper and other herbs to dilute alcohol prepared by distilling fermented maize or rye worts.

Malt whisky is prepared by fermentation and subsequent distillation of malted barley (having greater diastatic power than brewery malt), and grain whisky is prepared in a similar manner from a mixture of malted and unmalted barley with unmalted maize. These two forms of whisky when matured are finally blended to form Scotch whisky. Bourbon whisky is prepared from a mash containing maize as its main constituent and Irish whiskey from one containing predominantly rye.

5. Potentialities of micro-organisms for production of 2,3-butyleneglycol

Manufacture of synthetic rubber requires 1,3-butadiene, a key intermediate that can be readily obtained chemically from the mould metabolite 2,3-butyleneglycol. As might be expected, production of 2,3-butyleneglycol by fermentation was extensively investigated during the Second World War when the synthetic rubber industry was in its infancy. Much was learned of the fermentation; however, despite yields of 30–40% on the amount of carbohydrate consumed, the microbiological product cannot compete with 1,3-butadiene from petrochemical manufacture. It is possible that economic uses for the microbiological product may eventually be found in the fine chemicals industry. Meanwhile, description of the pilot plant fermentation will serve to illustrate another aspect of the versatility of micro-organisms in effecting chemical change.

Microbial 2,3-butyleneglycol occurs in three stereoisomeric forms: these are the laevo-rotatory or D-form, the dextro-rotatory or L-form, and the optically inactive meso-form represented by the following formulae:

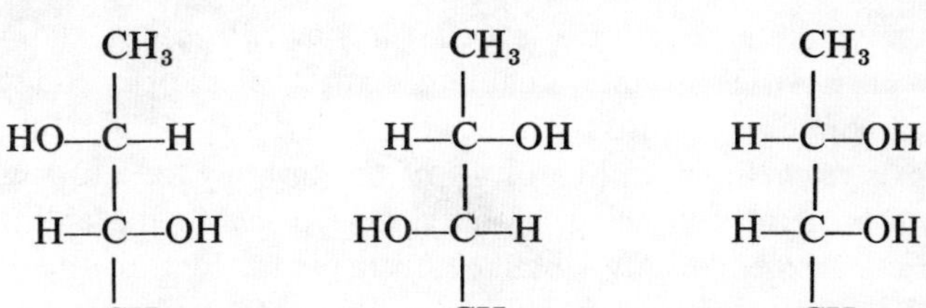

D(−)-2,3-butyleneglycol L(+)-2,3-butyleneglycol meso-2,3-butyleneglycol

2,3-Butyleneglycol can be formed microbiologically by two different pathways, as indicated in Fig. 10.2. Bacteria produce (+)-α-acetolactic acid by transferring active acetaldehyde to pyruvate, shown in route *E*, and then a stereospecific decarboxylation affords (—)-acetoin, but the (—)-acetoin of yeast is formed by transfer of active acetaldehyde to acetaldehyde by route *D*. The

(—)-acetoin formed by either metabolic path is converted by hydrogenation to 2,3-butyleneglycol. Dehydrogenation of (—)-acetoin to diacetyl is equally possible; it will therefore be clear that a low redox potential in the fermentation will be needed if 2,3-butyleneglycol is to be the end-product.

Various bacteria have been successfully used for producing 2,3-butyleneglycol by fermentations: *Bacillus polymyxa* produces only the D-form (laevo-rotatory), *Aenomonas hydrophila* and *Bacillus subtilis* produce mixtures of the D-form with the optically inactive meso-form, and *Serratia marcescens* affords a small amount of the L-form (dextro-rotatory) mixed with the predominant meso-form. The pattern of distribution of the stereoisomers is therefore closely similar to that already observed in lactic acid fermentations. The exact method of formation by bacteria of the three stereoisomeric forms of 2,3-butyleneglycol is in doubt: they may be derived from (—)-acetoin either by racemization with subsequent action of stereospecific 2,3-butyleneglycol dehydrogenases or, alternatively, racemization may not occur until the 2,3-butylene glycol has been formed.

The nature of the carbohydrate sources for the fermentation is not critical so long as they are readily fermentable; they must also be cheap. Some organisms (e.g. *Bacillus polymyxa*) are endowed with amylolytic enzymes, enabling them to make direct use of starch-containing substrates, but others lack polysaccharases and so involve preliminary hydrolysis of the carbohydrate to sugars. Many bacteria producing 2,3-butyleneglycol have the ability to utilize pentoses readily obtained from wood hydrolysates and sulphite waste liquor. Nitrogen and mineral contents of the medium are not critical, but supplementation with a source of thiamine is sometimes necessary.

The bacterial fermentation is begun at pH 6·2 approximately, and the medium needs to be buffered above pH 5·6 by inclusion of calcium carbonate to counteract the effects of organic acids produced by the micro-organism. It is notable that the (+)-α-acetolactic acid decarboxylase needed for (—)-acetoin formation has optimum activity at pH 6·0 and none at pH 7·8–8·0.

Temperatures within the range 30–37°C are optimal with sugar concentrations of up to about 8 % w/v; for anaerobic fermentation the medium is seeded with a non-critical concentration above 1 % v/v of a vigorously growing culture; the fermentation is usually completed within 2–3 days.

The bacterial route to 2,3-butyleneglycol is based on the same α-hydroxyethyl-TPP (active acetaldehyde) donor and the same acetaldehyde acceptor as that needed for ethanol production, and it is therefore not surprising to find that ethyl alcohol is a minor product of the anaerobic bacterial fermentation. It is to be expected that ethanol production could be held to a minimum by ensuring throughout the fermentation period an optimal pH (6·0) for operation of the stereospecific α-acetolactic acid decarboxylase.

The high boiling points (about 180°C) of mixtures of the 2,3-butyleneglycol stereoisomers present difficulties in recovery at the end of fermentation: ethanol is first removed, and the broth is then filtered and concentrated by evaporation before recovery of the required product by diverse means.

Suggested literature for further reading

1. Brian, P. W. and Radley, M. E., *Malting of Barley*, U.S.A. patent 3104207 (1963).
2. Bunker, H. J., Recent research on the yeasts, *Prog. ind. Microbiol.* **3,** 1–41 (1961).
3. Davis, J. G., The Lactobacilli—I, *Prog. ind. Microbiol.* **2,** 1–26 (1960).
4. Davis, J. G., The Lactobacilli—II. Applied aspects, *Prog. ind. Microbiol.* **4,** 95–136 (1962).
5. Dunn, C. G., The industrial utilization of yeasts, *Econ. Bot.* **12,** 145–63 (1958).
6. Gray, P. P., *Some Advances in Microbiological Control for Beer Quality, Wallerstein Advances in Beer Quality*, Part II, Wallerstein Laboratories Communications, **25,** (87) 166–79 (1962.)
7. Gray, P. P. *Some Practical Fermentation Problems in the Light of Recent Yeast Researches, Wallerstein Advances in Beer Quality*, Part II, Wallerstein Laboratories Communications, **25** (87) 180–9 (1962.)
8. Haas, G. J., Microbial control methods in the brewery, *Adv. appl. Microbiol.* **2,** 113–62 (1960).
9. Juni, E., Mechanisms of formation of acetoin by bacteria, *J. biol. Chem.* **195** (2) 715–26 (1952).

10. JUNI, E., Mechanisms of the formation of acetoin by yeast and mammalian tissue, *J. biol. Chem.* **195** (2) 727–34 (1952).
11. KOBAYASHI, Y. and KALNITSKY, G., The bacterial synthesis of α-acetolactate, *J. biol. Chem.* **211** (1) 473–82 (1954).
12. LONG, S. K. and PATRICK, R., The present status of the 2,3-butyleneglycol fermentation, *Adv. appl. Microbiol.* **5**, 135–55 (1963).
13. PREECE, J. A., *The Biochemistry of Brewing*, Oliver & Boyd, London, 1954.
14. RADHAKRISHNAN, A. N. and SNELL, E. E., Biosynthesis of valine and isoleucine. II. Formation of α-acetolactate and α-aceto-α-hydroxybutyrate in *Neurospora crassa and Escherichia coli, J. biol. Chem.* **235** (8) 2316–21 (1960).
15. ROSS, D., The acetone–butanol fermentation, *Prog. ind. Microbiol.* **3**, 71–90 (1961).
16. WALLERSTEIN, L., SALETAN, L. T. and GRAY, P. P., *A Review of Progress in Brewing Science and Technology, Wallerstein Advances in Beer Quality*, Part I, Wallerstein Laboratories Communications, **24** (84) 114–29 (1961.)
17. ATKIN, L. and GRAY, P. P., *Influence of Wort Constituents on the Rate of Fermentation, Wallerstein Advances in Beer Quality*, Part II, Wallerstein Laboratories Communications, **25** (87) 87–96 (1962.)

CHAPTER 11

Fermentation Products Formed by the Anaerobic Activities of Butyric Acid Bacteria

WE HAVE seen (Chapter 10) that oxidation of α-hydroxyethyl-TPP plays a key role in the oxidative decarboxylation of pyruvate, but that under varying degrees of anaerobiosis alternative metabolic pathways may be substituted to permit transfer of active acetaldehyde to suitable acceptors with formation of new compounds (Fig. 10.2). The processes occurring in these facultative anaerobes compare unfavourably with oxidative decarboxylation in efficiency of energy capture from the substrate. More efficient systems appear to be available to some obligate anaerobes, as exemplified by the metabolic activities of the anaerobic spore-forming rod-shaped bacteria of soil included in the genus *Clostridium* and collectively known as the butyric acid bacteria: these will now be considered.

1. Butanol–acetone fermentations

Clostridium species capable of anaerobic fermentation of polysaccharides, sugars and hydroxyacids to a variety of products, including butyric and acetic acids, n-butanol, acetone, ethanol, carbon dioxide and hydrogen, are widespread in nature. They have been used commercially for many years in the production of acetone and butanol, but the economics of these fermentation processes have become less favourable than they were. It is probable that acetone and butanol produced by fermentation are in an economically similar situation to that of fermentation alcohol:

countries with well developed petrochemical industries can produce the synthetic products more cheaply, but fermentation processes may remain economically viable in countries not so well developed, particularly when cheap raw materials are available for use in fermentation media. Butanol is the more valuable product, and the fermentation is directed towards maximum yield of this chemical.

Clostridium species are usually able to ferment sucrose, and various molasses may therefore be used as convenient raw materials for the fermentation media. Phosphorolysis of sucrose is succeeded by degradation of the resultant hexose phosphate to pyruvic acid, probably by means of the Embden–Meyerhof–Parnas glycolytic pathway (see Chapter 9), but further metabolism under anaerobic conditions then yields butyric acid and, finally, the end-products butanol and acetone.

1.1. *Formation of butyric acid*

Clostridium butylicum is able to decompose pyruvic acid to acetyl phosphate, with liberation of carbon dioxide and hydrogen, according to the equation

$$CH_3COCOOH + HOPO_3H_2 \longrightarrow CH_3COOPO_3H_2 + CO_2 + H_2$$

The reaction mechanism is probably complex, requiring TPP, coenzyme A and ferrous ions: acetyl-coenzyme A appears to be an intermediate, the final conversion to acetyl phosphate being mediated by the enzyme phosphotransacetylase:

$$\text{acetyl-coenzyme A} + \text{phosphate} \xrightleftharpoons{\text{phosphotransacetylase}} \text{acetyl phosphate} + \text{coenzyme A}$$

Formation of acetyl phosphate by this means may be a convenient method for storage of energy-rich acetyl groups or for generation of ATP by acetokinase action:

$$\text{acetyl phosphate} + \text{ADP} \xrightleftharpoons{\text{acetokinase}} \text{acetate} + \text{ATP}$$

It was explained in Chapter 10 that acetyl-coenzyme A is produced in aerobic organisms by oxidative decarboxylation of pyruvate through an intermediate α-hydroxyethyl-TPP and that this complex is immediately dehydrogenated by transfer of a hydrogen atom to α-lipoic acid, with subsequent serial transfers involving FAD, NAD and the enzyme chain of aerobic respiration to oxygen as the ultimate acceptor. This process cannot be operating in butyric acid bacteria because most of their fermentations produce hydrogen as a conspicuous product. The suggestion has been made that butyric acid bacteria are able, in some way, to modify the anaerobic reaction of yeasts that gives acetaldehyde (Fig. 10.2: route *C*), so that the hydrogen atom normally retained within the active acetaldehyde complex is now released as gaseous hydrogen. Whatever the detailed mechanism involved, it is clear that this process is energetically more efficient than alcoholic fermentation, because the energy-rich compound, acetyl-coenzyme A, is made available for subsequent use.

This key compound is also formed in a two-stage reaction from ethanol by *C. kluyveri*, a rather atypical member of the butyric acid bacteria that is unable to utilize carbohydrates or pyruvate, but can grow well on ethanol and acetate:

(i)

$$CH_3CH_2OH + NAD^{\oplus} \xrightleftharpoons[]{\text{alcohol dehydrogenase}} CH_3CHO + NADH + H^{\oplus}$$

(ii)

$$CH_3CHO + HS.Co\overline{A} + NAD^{\oplus} \xrightleftharpoons[]{\text{acetaldehyde dehydrogenase}} CH_3CO{\sim}S.Co\overline{A} + NADH + H^{\oplus}$$

Acetyl-coenzyme A, whether formed by *C. butylicum* from sugar or by *C. kluyveri* from ethanol, does not normally accumulate: it is usually consumed in biosynthetic reactions or stored as acetyl phosphate until required for generation of acetyl groups or ATP.

(i) acetyl-coenzyme A + phosphate

$$\xrightleftharpoons{\text{phosphotransacetylase}} \text{acetyl-phosphate} + \text{coenzyme A}$$

(ii) acetyl phosphate + ADP

$$\rightleftharpoons \text{acetate} + \text{ATP}$$

The main reaction consuming acetyl-coenzyme A in the butyric acid fermentation is biosynthesis of the key compound acetoacetyl-coenzyme A from two molecules of acetyl-coenzyme A:

$$2CH_3CO{\sim}S.Co\overline{A} \xrightleftharpoons{\text{thiolase}} CH_3COCH_2CO{\sim}S.Co\overline{A} + HS.Co\overline{A}$$

Acetoacetyl-coenzyme A is reduced by the enzyme, β-hydroxybutyryl-coenzyme A dehydrogenase (working in reverse) to β-hydroxybutyryl-coenzyme A. Dehydration by the enzyme crotonase yields crotonyl-coenzyme A, which is reduced by the enzyme butyryl-coenzyme A dehydrogenase (working in reverse) to butyryl-coenzyme A.

1.2. *Formation of butanol*

Butyryl-coenzyme A is enzymically reduced, probably by enzymes similar to the aldehyde dehydrogenase of *C. kluyveri*, to butyraldehyde and then further by alcohol dehydrogenase to butanol:

(i) $$CH_3CH_2CH_2CO{\sim}S.Co\overline{A} + NADH + H^{\oplus} \rightleftharpoons CH_3CH_2CH_2CHO + HS.Co\overline{A} + NAD^{\oplus}$$

(ii) $$CH_3CH_2CH_2CHO + NADH + H^{\oplus} \xrightleftharpoons{\text{alcohol dehydrogenase}} CH_3CH_2CH_2CH_2OH + NAD^{\oplus}$$

1.3. *Formation of acetone*

The key compound, acetoacetyl-coenzyme A, in the biosynthesis of butyric acid is usually diverted to a certain extent to yield acetoacetic acid, which is then irreversibly decarboxylated to acetone. It is probable that the thioester bond energy of aceto-acetyl-coenzyme A is conserved during this process by substitution of a transfer reaction for the simpler but wasteful hydrolytic cleavage of the thioester bond:

(i)

$$CH_3COCH_2CO{\sim}S.Co\overline{A}+CH_3COOH \xrightleftharpoons{\text{transferase}} CH_3COCH_2COOH+CH_3CO{\sim}S.Co\overline{A}$$

(ii)

$$CH_3COCH_2COOH \xrightleftharpoons{\substack{\text{acetoacetate}\\ \text{decarboxylase}}} CH_3COCH_3+CO_2$$

1.4. *Interrelations between the biosynthetic pathways to butanol and to acetone*

Oxidation of sugar by the Embden–Meyerhof–Parnas glycolytic pathway is accompanied by reduction of NAD. Continuation of glycolysis requires continual regeneration of the oxidized form of NAD, and this is usually achieved by dehydrogenation with FAD and electron transfer to suitable acceptor substances. Two electron-accepting reactions are dominant when the fermentation is producing butyric acid: these are the reduction of acetoacetyl-coenzyme A to β-hydroxybutyryl-coenzyme A and the reduction of crotonyl-coenzyme A to butyryl-coenzyme A. The side-reaction whereby acetoacetyl-coenzyme A is converted to acetone results in loss of both of these electron-accepting reactions: continued glycolysis demands that these should be replaced by alternative electron-accepting reactions. Reduction of butyryl-coenzyme A to butyraldehyde and then to butanol is a convenient solution. It is probable that analogous reactions involving reduction of acetyl-coenzyme A to acetaldehyde and ethanol play supporting roles. The outcome of this metabolic shunt is that production of both butanol and ethanol is a necessary corollary of acetone formation in the fermentation, the key reaction being the decarboxylation of acetoacetate.

1.5. *The course of the fermentation*

Selected strains of *C. acetobutylicum* are usually used for industrial fermentations: the organisms are spore-forming rods, may conveniently be isolated from soil under anaerobic conditions and may be stored in dry sterile soil for long periods.

A culture of the selected organism that has been grown and allowed to sporulate in a corn- or potato-based medium containing about 4% w/v starch is usually heated at about 65–100°C for up to 3 min, to kill the weaker spores before inoculation into a larger volume of similar medium maintained under anaerobic conditions. Rapid growth in the inoculum development medium is permitted for about 24 hr before transfer of about 0·5–5·0% v/v of inoculum, in the logarithmic phase of growth, to the fermentation medium maintained close to the optimum temperature of 30–32°C: anaerobic conditions are achieved by maintaining over the culture a positive pressure of filtered gas from other fermentations.

Various raw materials may be used in the fermentation medium. Suitable carbon sources, depending upon capacity for utilization by the selected organism, include invert, blackstrap, beet or citrus molasses, sulphite waste liquor, inverted starches, and starch-containing grains: they are usually employed to give a concentration equivalent to 5–7% w/v hexose and the pH is adjusted so as to give about 6·0 to 6·5 after sterilization.

Ammonium hydroxide or ammonia is usually added at intervals from the beginning of fermentation to serve as a source of nitrogen and to counteract the tendency towards excessive acidity by maintaining the pH above 5·5. Some strains require supplementation of the nitrogen source with degraded protein for best results, and this may be provided as cornsteep liquor or as yeast autolysate. A few strains require biotin and *para*-aminobenzoic acid as accessory food factors.

During the early and middle period of fermentation carbon dioxide and hydrogen are evolved in the manner already discussed, and this helps to maintain anaerobiosis. The organic acids, mainly butyric, formed during this stage of fermentation tend to depress the pH, but their effect is counteracted by further additions of ammonium hydroxide or ammonia.

Solvent formation occurs relatively late in the fermentation, when the tendency for pH depression is greatest, and this is associated with a rapid increase in concentration of acetoacetate decarboxylase and then by a diminution in concentration of volatile acids as the fermentation progresses.

The fermentation is usually completed within 48–60 hr, when the mixed solvent concentration (acetone/butanol/ethanol mixture) in the broth may be about 2% v/v. The proportion of butanol:acetone:ethanol is usually around 6:3:1, but this depends to some extent on the strain selected and the conditions of fermentation: a somewhat higher proportion of the more valuable butanol has been reported. Maximum yields claimed are of the order of 30% mixed solvent on the weight of fermentable carbohydrate supplied.

Contaminants are relatively few, but lactic acid bacteria that are able to grow under the micro-aerophilic conditions employed may be troublesome. Infection may usually be readily detected by diminished evolution of gas or by excessive acid production. Infection of broths by bacteriophages (bacterial viruses) may sometimes occur and result in death of the fermenting organism. Bacteriophage infection can only be controlled by use of immune strains, but these may be readily obtained by application of standard microbiological methods.

The harvested broth is stored for a period in a holding tank before recovery of the acetone and butanol by fractional distillation.

1.6. *By-products of the fermentation*

The gas evolved in considerable quantity during the fermentation contains carbon dioxide and hydrogen in the approximate ratio of 2:1. The carbon dioxide may be converted to solid carbon dioxide and the hydrogen is sometimes used in industrial chemical syntheses. Attempts have also been made to concentrate and then dry the stillage (composed largely of bacterial cell residues) for use in animal feeds as a high protein supplement containing an appreciable amount of riboflavin.

Suggested literature for further reading

1. BARKER, H. A., The chemistry of butyric acid–butanol fermentations (Chapter 2 in *Bacterial Fermentations*), Ciba lectures in microbial biochemistry, Wiley, New York, 1956.
2. MCCLUNG, L. S., The anaerobic bacteria with special reference to the genus *Clostridium*, *A. Rev. Microbiol.* **10,** 173–92 (1956).
3. ROSS, D., The acetone–butanol fermentation, *Prog. ind. Microbiol.* **3,** 71–90 (1961).

CHAPTER 12

Fermentation Products Related to Intermediates of the Citrate Cycle

GLYCOLYSIS of sugars to pyruvic acid and oxidative decarboxylation under aerobic conditions of pyruvate to acetyl-coenzyme A and carbon dioxide was given in Chapters 9-10 it was noted that complete oxidation to carbon dioxide and water then commonly occurred.

Complete oxidation of acetate is effected by a group of coordinated, enzymic processes, collectively termed the citrate cycle, which normally works in conjunction with the respiratory chain of enzymes:

$$CH_3COOH+4O = 2CO_2+2H_2O \qquad (12.1)$$

1. The citrate cycle (see Fig. 12.1)

Acetate enters the citrate cycle as acetyl-coenzyme A and condenses with the enol form of oxaloacetic acid to give citric acid. Isomerization to isocitric acid occurs via the intermediate *cis*-aconitic acid. Isocitric acid is dehydrogenated to oxalosuccinic acid, and this intermediate is decarboxylated to give α-keto-glutaric acid. Oxidative decarboxylation of α-ketoglutaric acid to succinyl-coenzyme A or active succinate then occurs in the manner typical of α-keto acids (see Chapter 10). Succinyl-coenzyme A is converted to succinic acid, but the bond energy is preserved by coupling to another reaction that transforms guanosine diphosphate (GDP) and orthophosphate (Pi) to energy-rich guanosine triphosphate (GTP). Succinic acid is dehydrogenated to give fumaric acid, and subsequent hydration affords malic acid. Another dehydrogenation completes the cycle by converting malic acid to oxaloacetic acid. It should be noted that all the carboxylic acids of the citrate cycle occur in salt form at physiological pH.

In this cycle, oxaloacetic acid acts as a carrier for activated acetate, permitting complete oxidation of 1 mole in each complete cycle without loss of the C_4 compound (see eq. (12.1)). No oxygen is taken up in this process. The four oxygen atoms shown in eq. (12.1) correspond to the four pairs of hydrogen atoms removed by the citrate cycle dehydrogenations, because the hydrogen acceptors (NAD, NADP and FAD) are simultaneously reduced and their reduced forms are subsequently re-oxidized by atmospheric oxygen through the mediation of the respiratory chain of enzymes, with concomitant but stepwise formation of water. The overall process is the main energy-producing reaction available to biological systems, the change in free energy being only a little less than when water is formed from its constituent elements:

$$2H_2 + O_2 = 2H_2O \; (\Delta F = 57 \text{ kcal/mole } H_2O)$$

Molecular oxygen cannot oxidize directly the reduced forms of the hydrogen acceptors. Nevertheless, the oxidation (loss of electrons ($e^{\ominus}$)) of the pairs of enzyme-bound hydrogen atoms by an oxidizing agent (electron acceptor) can be coupled to the ultimate oxidizing agent, molecular oxygen, with overall formation of water:

(i) $$\underset{\text{(enzyme-bound)}}{2H} - 2e^{\ominus} = 2H^{\oplus}$$

(ii) $$\tfrac{1}{2}O_2 + 2e^{\ominus} = O^{\ominus\ominus}$$

(iii) $$O^{\ominus\ominus} + 2H^{\oplus} = H_2O$$

Overall: $$\underset{\text{(enzyme-bound)}}{2H} + \tfrac{1}{2}O_2 = H_2O$$

Coupling of reaction (i) to reaction (ii) must be done stepwise cause of the relatively large change of free energy: intermediate electron acceptors are therefore necessary, and these can operate only in a sequence of ascending oxidizing capability (i.e. successively higher rH values). Oxidation of a substrate SH_2 by atmospheric oxygen with stepwise formation of water in a biological system ordinarily conforms to the general pattern shown in Fig. 12.2.

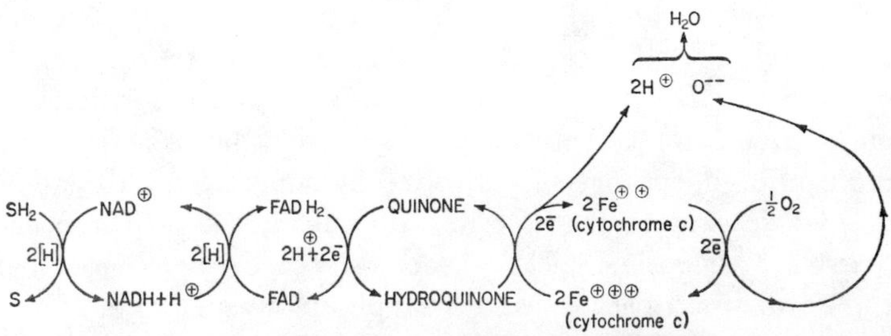

FIG. 12.2. The respiratory chain of enzymes.

The enzymes concerned are collectively termed the respiratory chain of enzymes: they constitute a coupled series of redox systems, permitting conservation of some of the free energy of water formation by subdividing the overall exergonic reaction into a series of smaller energy steps that can be individually coupled with processes for generation of ATP (collectively termed oxidative phosphorylation). The mechanism of oxidative phosphorylation has not been fully determined.

Just as the dehydrogenations occurring in the citrate cycle are linked to metabolic energy-conserving processes in the respiratory enzyme chain, so the intermediates of the cycle are associated with other metabolic processes: these are both degradative and biosynthetic. The citrate cycle thus furnishes a pool of interchangeable intermediates linking the diverse pathways concerned with metabolism of carbohydrate, fat and protein.

1.1. *Citrate cycle links with carbohydrate*

Degradation of sugars with conservation of much of the chemical energy proceeds through the Embden–Meyerhof–Parnas glycolytic pathway to pyruvate (see Chapter 9) and then, under aerobic conditions, by oxidative decarboxylation to acetyl-coenzyme A, thereby permitting entry to and complete oxidation in the citrate cycle. Alternative mechanisms permitting continued glycolysis under temporary or complete anaerobiosis have been discussed in Chapters 10 and 11.

The glycolytic reactions are theoretically reversible, but the equilibrium of the pyruvate kinase reaction (phosphoenol-pyruvate + ADP $\rightleftharpoons$ pyruvate + ATP), in particular, is unfavourable for resynthesis of sugars (gluconeogenesis). The problem of gluconeogenesis is solved by reductive carboxylation of pyruvic acid to malic acid, with subsequent dehydrogenation to oxaloacetic acid (Wood–Werkman reaction): the enol-form of oxaloacetic acid is then phosphorylated to phosphoenolpyruvic acid by inosine triphosphate (ITP) with concomitant decarboxylation (Fig. 12.1).

1.2. *Citrate cycle links with amino acids*

Many amino acids are termed glucogenic because they can be metabolized to glucose. The citrate cycle is an intermediary in this conversion, but it also provides a mechanism for the complete degradation of these amino acids. The linkages are provided by various intermediates.

(i) Oxaloacetic acid (reversible transamination to give aspartic acid).
(ii) α-Ketoglutaric acid (reversible reductive amination to give glutamic acid).
(iii) Succinyl-coenzyme A (formed from propionyl-coenzyme A via methylmalonyl-coenzyme A).

(iv) Fumaric acid (formed from the aspartic acid associated with the urea cycle, and as one of the products of degradation of the ketogenic acids, tyrosine and phenylalanine).

Further metabolic links with other amino acids then become possible, because glutamic acid can be formed from arginine and ornithine (urea cycle amino acids), from histidine and from proline; and propionyl-coenzyme A is a product of valine degradation.

It should also be noted that alanine can enter the citrate cycle as acetyl-coenzyme A after preliminary transamination to pyruvic acid, and that this provides a route for degradation of any amino acid that can give alanine: these include serine, glycine and cysteine.

Metabolic products formed from other amino acids (phenylalanine, leucine, isoleucine and tyrosine) include acetoacetate (one of the "ketone bodies"), and these amino acids are termed "ketogenic" for this reason. The ketogenic amino acids are linked to the citrate cycle via acetyl-coenzyme A in the same way as acetyl-coenzyme A links the fatty acids derived from fats.

1.3. *Citrate cycle links with fatty acids*

The long-chain fatty acids having an even number of carbon atoms derived from hydrolysis of fats are activated by formation of the acyl-coenzyme A derivative (this requires ATP and combination with coenzyme A). The energy-rich thioester derivative of the fatty acid can then be completely degraded to acetyl-coenzyme A by successive β-oxidations, thereby permitting entry to the citrate cycle. The process of β-oxidation involves a sequence of dehydrogenation, addition of water and further dehydrogenation to a β-keto acid. Thioclastic cleavage of a C—C bond in the presence of coenzyme A then removes the two terminal carbon atoms in the form of acetyl-coenzyme A, leaving a residual acyl-coenzyme A molecule that is primed ready to undergo another β-oxidation:

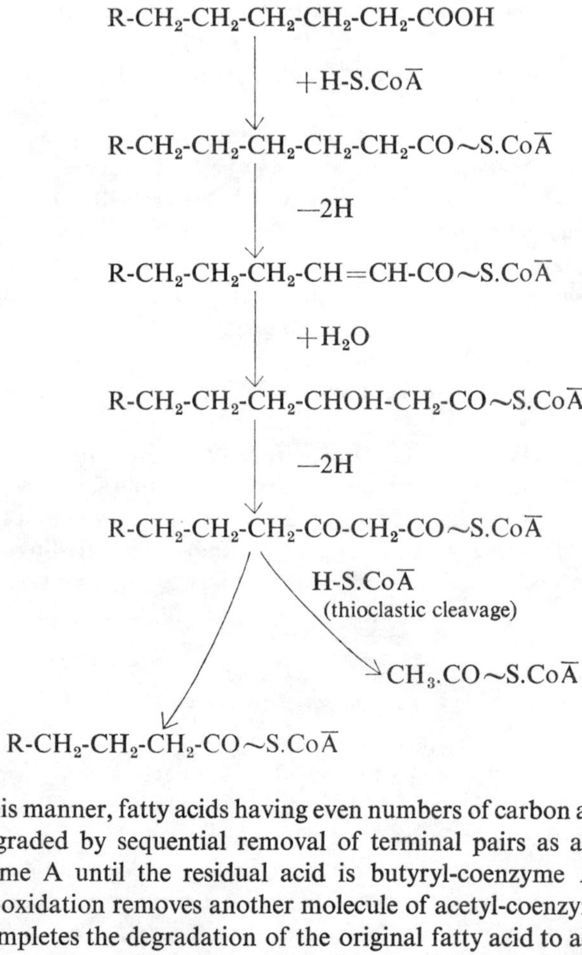

In this manner, fatty acids having even numbers of carbon atoms are degraded by sequential removal of terminal pairs as acetyl-coenzyme A until the residual acid is butyryl-coenzyme A. A final β-oxidation removes another molecule of acetyl-coenzyme A and completes the degradation of the original fatty acid to acetyl-coenzyme A.

Unbranched fatty acids having an odd number of carbon atoms undergo β-oxidation in similar manner, but the residual acid is now propionyl-coenzyme A: this compound then enters the citrate cycle by conversion to succinyl-coenzyme A via the intermediate malonyl-coenzyme A, as already described.

1.4. *Citrate cycle links with tetrapyrroles*

Biosynthesis of the tetrapyrrole derivative, vitamin B_{12}, will be discussed in Chapter 15, and it is necessary here only to note that the citrate cycle is linked to tetrapyrroles by succinyl-coenzyme A.

2. Inhibitors of the citrate cycle

The citrate cycle cannot function when oxaloacetic acid or any other of the dicarboxylic acid intermediates is deficient. The Wood–Werkman reaction, when operative, protects against this eventuality by the reductive carboxylation of pyruvic acid to malic acid with dehydrogenation of the latter to oxaloacetic acid: the overall reaction effects a gain to the citrate cycle of one molecule of C_4 acid by formation of oxaloacetic acid from pyruvic acid and carbon dioxide.

Operation of the citrate cycle depends upon continuous provision of inorganic phosphate for the phosphorylation of GDP to GTP in the reaction that converts succinyl-coenzyme A to succinic acid. Phosphate is also consumed in the oxidative phosphorylation reactions associated with the respiratory chain of enzymes that bring about the stepwise oxidation of the four pairs of hydrogen atoms removed in the citrate cycle dehydrogenations.

Some of the reactions of the citrate cycle require, as prosthetic groups, metal ions in a suitable state of oxidation. Oxalosuccinic decarboxylase and the system for oxidative decarboxylation of α-ketoglutaric acid both need the presence of $Mn^{\oplus\oplus}$ or $Mg^{\oplus\oplus}$. Succinic dehydrogenase is a metallo-flavoprotein containing iron: the flavin group, similar to FAD, is reduced when succinic acid is dehydrogenated to fumaric acid, and the reduced coenzyme is re-oxidized by a number of intermediate carriers with ultimate donation of electrons to the ferric ion groups of cytochrome *c*:

$$Fe^{\oplus\oplus\oplus}+H \longrightarrow Fe^{\oplus\oplus}+H^{\oplus}$$
$$[Fe^{\oplus\oplus\oplus}+e^{\ominus} \longrightarrow Fe^{\oplus\oplus}]$$

Many of the enzymic reactions of the citrate cycle are susceptible to direct inhibition: aconitase is inhibited by fluorocitrate;

isocitrate dehydrogenase is inhibited by citric acid and by ferrocyanide; succinic dehydrogenase is inhibited by malonate; the oxidative decarboxylation of α-ketoglutaric acid is inhibited by arsenite. All the dehydrogenation reactions are ultimately dependent upon maintenance of aerobic conditions, but isocitric dehydrogenase is particularly susceptible to inhibition by anaerobiosis.

3. The glyoxylate cycle

Some micro-organisms are able to use a variation of the citrate cycle that has been named the glyoxylate cycle.

The function of the glyoxylate cycle is to synthesize C_4-dicarboxylic acids from acetyl-coenzyme A and thereby permit ultimate synthesis of carbohydrate by an alternative mechanism to the citrate cycle. The glyoxylate cycle plays a major part in the metabolism of micro-organisms able to use fatty acids as sole source of carbon by degradation to C_2 compounds. The enzymes comprising the glyoxylate cycle appear to be inducible enzymes (see Chapter 8) formed when carbohydrate is replaced by acetate or its nutritional equivalent.

The glyoxylate cycle is summarized in Fig. 12.3. The early stages follow the course of the citrate cycle: acetyl-coenzyme A condenses with the enol form of oxaloacetic acid to give citric acid, which is isomerized to isocitric acid. At this stage an inducible enzyme, isocitritase, intervenes to split isocitric acid into glyoxylic acid and succinic acid. The glyoxylic acid continues in the glyoxylate cycle by combining with another molecule of acetyl-coenzyme A to give malic acid, which is converted to oxaloacetic acid as in the citrate cycle.

The effect of each complete glyoxylate cycle is to form one mole of succinic acid from two moles of acetyl-coenzyme A and to regenerate the mole of oxaloacetic acid needed to begin another cycle. In other words, the glyoxylate cycle employs glyoxylic acid and an additional mole of acetyl-coenzyme A to short-circuit the citrate cycle between isocitric acid and malic acid, with con-

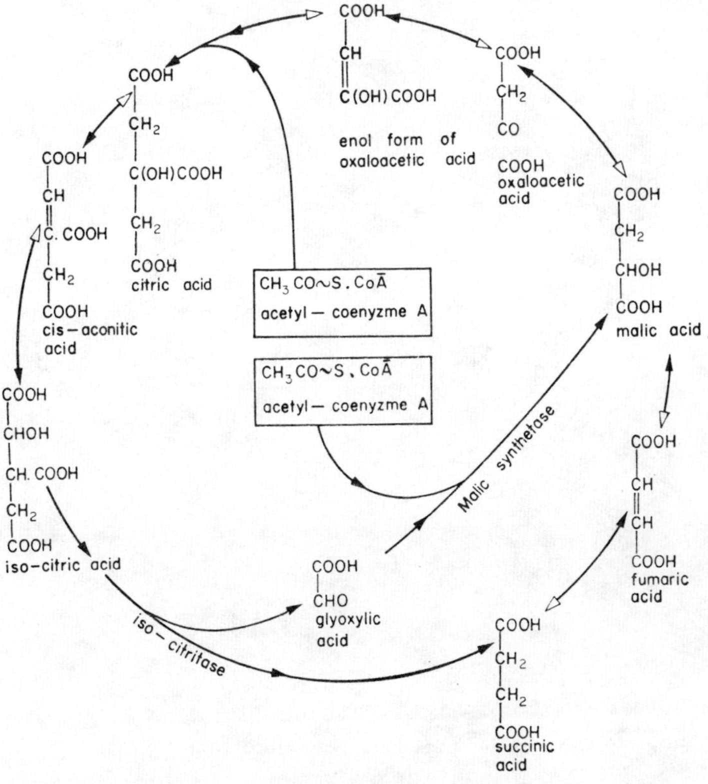

FIG. 12.3. The glyoxylate cycle.

comitant gain of one mole of succinic acid. The glyoxylate cycle may also serve to produce glyoxylic acid for biosynthesis of the amino acid glycine.

4. Production of organic acids

Organic acids related to the citrate cycle produced by industrial fermentation processes include citric acid, fumaric acid and glutamic acid (from α-ketoglutaric acid). The mechanism

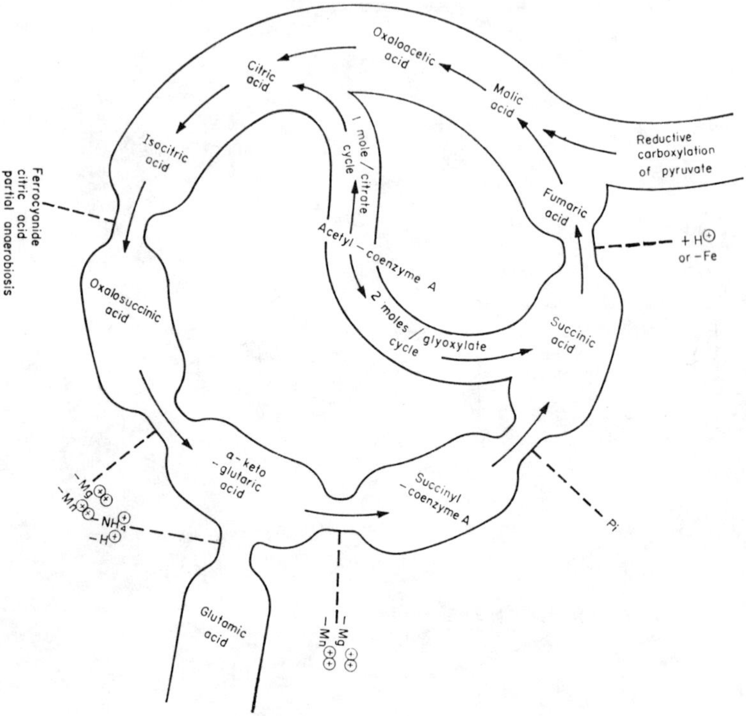

FIG. 12.4. Metabolic "pipe-lines" concerned with production of organic acids.

for production of these organic acids may conveniently be considered as the system of pipe-lines shown in Fig. 12.4. The main anticlockwise channel corresponds to the citrate cycle, but its bore may be considered to be constricted at specific points by unfavourable metabolic conditions.

(i) Flow from isocitric acid to oxalosuccinic acid is retarded by presence of free citric acid and by ferrocyanide.

(ii) Flow from oxalosuccinic acid to α-ketoglutaric acid is diminished by absence of $Mn^{\oplus\oplus}$ or of $Mg^{\oplus\oplus}$.

(iii) Channel for oxidative decarboxylation of α-ketoglutaric acid to succinyl-coenzyme A is constricted by absence of $Mn^{\oplus\oplus}$ or $Mg^{\oplus\oplus}$.

(iv) Flow of succinyl-coenzyme A to succinic acid is diminished by a limiting concentration of inorganic phosphate.

(v) Bore of the pipe conveying succinic acid into fumaric acid is diminished by a high hydrogen ion concentration or by absence of iron, both interfering with functioning of succinic dehydrogenase, as already explained.

The main citrate cycle pipe-line is fed by an inlet providing acetyl-coenzyme A. This system is unable to continue accumulating any of the citrate cycle intermediates when the pipe is constricted at any of the points theoretically possible, because the cycle makes no gain of C_4 acid. The feeder-line able to supply malic acid, formed by reductive carboxylation of pyruvic acid, offers one solution to the problem: the feeder-line supplying succinic acid formed from 2 moles of acetyl-coenzyme A in the glyoxylate cycle offers another solution. In either event, the continuous gain of C_4 acid now permits accumulation of specific citrate cycle intermediates when the main pipe-line is constricted at appropriate positions.

4.1. *Citric acid*

Production of citric acid requires a medium that is initially highly acidic, and it is usual to begin at pH3 or less. An even lower pH is developed during fermentation, and it is necessary to employ acid-resistant materials, such as good quality stainless steel, aluminium or ceramics, for the fermentation equipment.

Almost theoretical yields of citric acid can be obtained from cheap sources of carbohydrate by fermentation with special strains of *Aspergillus niger*: both surface cultures and submerged fermentation processes are employed. The concentrations of phosphate, iron, manganese, magnesium and zinc must be limited, in order to restrict the flow in the citrate cycle (see Figs. 12.1 and

12.4): for this purpose, the acid medium is treated either with ferrocyanide to bind ions or with ion exchangers. Further restrictions are imposed on the citrate cycle by inhibition of isocitric dehydrogenase: this enzyme is susceptible to inhibition by anaerobiosis, so that limited aeration will aid citric acid production, but ferrocyanide incorporated in the medium has the additional function of acting as a direct inhibitor of the enzyme. The product, citric acid itself, is an inhibitor of isocitric dehydrogenase, so that maintenance of low pH and avoidance of citric acid precipitation will also assist inhibition of the enzyme and lead to greater accumulation of citric acid.

There is a tendency to produce oxalic acid if the pH becomes neutral or alkaline (presumably formed from surplus glyoxylic acid of the glyoxylate cycle), but its formation can be almost totally suppressed by suitably low pH and by low concentrations of inorganic ions. Similarly, the fermentation may be misdirected if the concentration of available nitrogen is in excess: this produces low titratable acid in the broth, but gives a high weight of mycelium presumably because reductive amination of α-ketoglutaric acid to glutamic acid, a key compound in amino acid synthesis, is facilitated.

Production of citric acid in the surface culture process begins when spores of *A. niger* are blown on to the surface of the nutrient medium 3–4 in. deep in aluminium trays. The cultures are incubated at 24–30°C for about 10 days, depending on the strain employed, in fermentation chambers supplied with sterile air at the rate of one volume of air to each volume of medium in every 5 min period.

During the fermentation the spores germinate to form mycelium, and within 2–5 days the surface of the liquid is covered by the mycelial mat; the pH falls from 2·5 to 1·5 and titratable acidity rises to a maximum at harvest. Maximum growth is attained one day before harvest.

The composition of the medium is the key to the fermentation, but this will depend to some extent on the strain used. In general, the carbohydrate can be glucose (which may yield as much as 90%

citric acid), sucrose, fructose, maltose or crude sugars (beet or cane molasses), and sugar concentrations between 14% and 20% w/v may be used. Citric acid can also be produced by some strains from polysaccharides and from acetic acid (usually used as the calcium salt). It may be that some acetate or its metabolic equivalent may have a beneficial effect in inducing activity of the enzymes of the glyoxylate cycle. The nitrogen source may be inorganic (e.g. ammonium nitrate) or organic (e.g. molasses), but to limit growth the concentration of available nitrogen should not exceed 0·08–0·09% w/v.

The concentrations of the other inorganic ions must also be carefully controlled. If the fermentation is to be phosphate-limited, the phosphoric acid added to the medium must not raise the total concentration above the equivalent of 0·03% w/v KH_2PO_4. If the fermentation is to be directed by limitation of the iron content of the medium, then potassium ferrocyanide is added until the iron concentration of the medium falls below 0·03% w/v. This treatment probably reduces the concentration of other cations, including zinc, that would tend to stimulate growth of mycelium at the expense of acid production.

Maintenance of sterile conditions in the fermentation broth is no problem, because of the low pH, but the fermentation medium is always heated to 100°C or more before inoculation, and the influent air is filtered through cotton-wool pads as an aid to maintaining aseptic conditions. The fermentation chambers are constructed to permit thorough cleansing and sterilization with 6% formaldehyde. Gross infection is removed manually, either by rejection of one complete tray or by excision of the infected area of the mat.

The conversion of carbohydrate to citric acid depends on intracellular enzymes, so that sugar from the nutrient fluid must be able to enter the cells of the mycelium and the citric acid produced must be able to diffuse out into the liquid. For this reason no advantage is to be gained by making the depth of the fermentation medium more than 3–4 in.

At harvest the citric acid as the calcium salt is precipitated from the fermentation liquor and from washes of the mycelium by treatment between 80–90°C at neutral pH for 3–4 hr. The salt is removed by filtration before reconversion to citric acid by adding excess sulphuric acid. The crude citric acid is decolorized by activated charcoal, and the soluble iron impurities are removed by adding ferrocyanide. Calcium sulphate is removed by filtration through a filter press and the liquor is concentrated under vacuum to about 80% w/v before crystallization in cooled agitated crystallizers.

Two other methods for the fermentation of citric acid have been reported. One of these is a typical submerged culture fermentation in the same medium and with the same fermentation characteristics as that of the surface culture process: the pH drops rapidly during the first day to 2·0 and then more slowly to reach 1·5 at the harvest time of about 10 days. During this period the titratable acid increases to a maximum of about 70% w/w of the added sugar. Growth reaches a maximum 3–4 days before harvest.

The mycelium of *A. niger* in submerged culture may be in the form of pellets 1–2 mm diameter, and the efficiency of oxygen transfer to the mould under any standard conditions of agitation and aeration is therefore less than to a culture containing more uniformly distributed mycelium. This may be an important factor in the inhibition of isocitric dehydrogenase, a condition required for good yields of citric acid.

The third method holds mycelium pellets of *A. niger* in an unstirred tower containing the nutrient medium. Air enters the bottom of the tower, to be dispersed by a number of perforated plates or aeration discs situated at a higher level. Aeration efficiency is a function of the rate of air-flow, and it becomes maximal just before the system is flooded with air (see Chapter 7).

4.2. L-*Glutamic acid*

This amino acid plays multiple roles in microbial metabolism, with the result that it does not normally accumulate. Nevertheless, during recent years work in Japan has revealed that certain strains of coryneform bacteria (short, pleomorphic gram-positive non-sporing rods that are not motile and are devoid of flagellae) are able to produce large amounts (exceeding 30 g/l.) of L-glutamic acid in culture. These micro-organisms have been collectively named glutamic acid bacteria, and they have been used for industrial production of the L-isomer.

L-glutamic acid is formed from α-ketoglutaric acid, but its biosynthesis from carbohydrate involves many stages: carbohydrate must undergo aerobic oxidation to α-ketoglutaric acid via the citrate cycle, but a reductive amination, catalysed by glutamic acid dehydrogenase, is then necessary to complete the biosynthesis:

$$\mathrm{COOH.CH_2.CH_2.CO.COOH + NH_4^{\oplus} + NADPH\ (or\ NADH) + H^{\oplus} \underset{\text{L-glutamic acid dehydrogenase}}{\rightleftharpoons} COOH.CH_2.CH_2.CHNH_2.COOH + H_2O + NADP^{\oplus}\ (or\ NAD^{\oplus})}$$

The reductive amination does not necessarily require anaerobic conditions, but it does need a redox potential sufficiently low to supply $NADPH_2$ or $NADH_2$ in the presence of ammonium ion. The reaction simultaneously provides a mechanism for re-oxidation of $NADH+H^{\oplus}$ to the oxidized form $NAD^{\oplus}$ needed for oxidation of isocitric acid in the citrate cycle.

Numerous species have been claimed of value for industrial production of the L-glutamic acid required to meet the demand for monosodium glutamate as a flavour-enhancing food additive: these include strains of *Brevibacterium flavum*, *Bb. divaricatum*, *Bb. roseum*, *Bb. lactofermentus*, *Corynebacterium acetoacidophylum*, *Cb. callunae*, *Cb. lilium* and an organism originally described as *Micrococcus glutamicus*.

The glutamic acid bacteria usually exhibit a nutritional requirement for the accessory food factor, biotin, but this substance also aids in directing the fermentation towards production of L-glutamic acid, particularly when carbohydrates are used as carbon source. The mechanism of the action of biotin is not known, but it is believed to regulate cell wall permeability, perhaps by affecting the biosynthesis of unsaturated fatty acids.

Biotin is one of the constituent "bios factors" of "yeast growth factor", having the structure

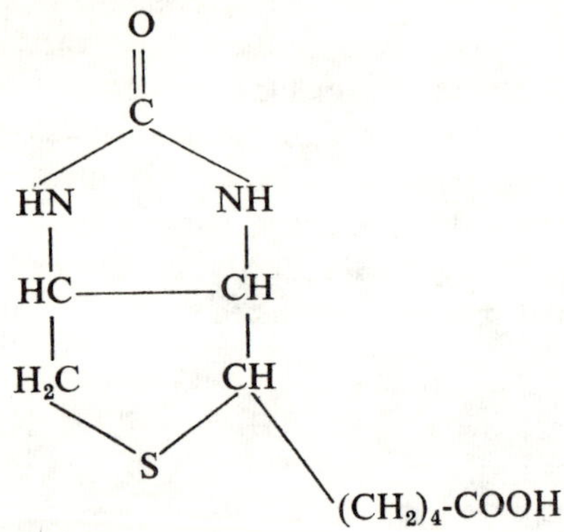

Biotin is the prosthetic group of enzymes responsible for many carboxylation reactions and its role in the formation of one such enzyme, propionyl-coenzyme A carboxylase, has been recently determined. Under the influence of holocarboxylase synthetase (apocarboxylase-activating enzyme), ATP and biotin are converted, in the presence of $Mg^{\oplus\oplus}$ to D-biotinyl-5′-adenylate:

$$\text{Biotin} + \text{ATP} \underset{}{\overset{Mg^{\oplus\oplus}}{\rightleftharpoons}} \text{biotinyl-AMP} + \text{PP}_1$$

The same enzyme then transfers biotin from biotinyl-AMP to the apocarboxylase forming propionyl-coenzyme A carboxylase and AMP:

$$\text{Biotinyl-AMP} + \text{apocarboxylase} \longrightarrow \text{carboxylase} + \text{AMP}$$

Cultures of glutamic acid bacteria are maintained on slopes of tryptone-glucose-yeast extract agar. Inoculum is developed in submerged aerated culture for 15–30 hr at 28–30°C in the same medium, with serial transfers if necessary to produce an adequate bulk of inoculum.

A good growth medium is required for fermentation, and this is usually provided by readily utilized sources of carbon (e.g. 5–20% of a suitable carbohydrate) and of nitrogen (e.g. urea, ammonium salts, cornsteep liquor or digested soya bean meal in an amount exceeding that required for theoretical conversion of the carbohydrate to glutamic acid), inorganic salts of calcium, potassium and magnesium, together with phosphate and sulphate ions and trace quantities of magnesium, manganese, iron, zinc and cobalt and biotin.

The amount of biotin supplied in the medium is adjusted according to the carbohydrate concentration: optimum concentrations are approximately 1·5–2 μg/l. for 5% sugar, 2–3 μg/l. for 10% sugar, 4–8 μg/l. for 15% sugar and 7–10 μg/l. for 20% sugar, and allowance should be made for the biotin content of any complex medium ingredients (see Chapter 6).

It should be noted, however, that some strains of glutamic acid bacteria are able to give good yields of L-glutamic acid from acetic acid (usually supplied in the form of Na or NH_4 salts) as carbon source.

The usual type of carbohydrate based medium is adjusted to pH 6–8 before inoculation with 2–10% v/v of vigorously growing inoculum. The temperature is usually maintained between 27–30°C with fairly good aeration, and it is essential to maintain pH 6·0–8·0 with ammonium ions continuously available in order to swing the equilibrium in favour of the reductive amination of α-ketoglutaric acid.

Microsporum glutamicus and probably other glutamic acid bacteria operate both the Embden–Meyerhof–Parnas and the Warburg–Dickens oxidative pentose-phosphate pathways for glucose dissimilation (see Chapter 9). Given good aeration, the pentose-phosphate oxidative pathway is favoured and this increases the yield of L-glutamic acid. Anaerobic conditions, on the other hand, favour the Embden–Meyerhof–Parnas pathway, with the result that the yield of glutamic acid is less favourable, because lactic acid (see Chapter 10) formed from pyruvic acid tends to accumulate instead.

The microbial culture grows vigorously in the fermentation medium for the first 10–20 hr. The growth rate then diminishes, and the main accumulation of L-glutamic acid begins. The fermentation is terminated at the end of about 30–50 hr, when the L-glutamic acid concentration has reached a maximum.

Suggested literature for further reading

1. Asai, T., Aida, K. and Oishi, K., L-glutamic acid by fermentation, *Bull. agric. chem. Soc. Japan* **21,** 134–5 (1957).
2. Christner, J. E., Schlesinger, M. J. and Coon, M. J., Enzymatic activation of biotin: biotinyl adenylate formation, *J. biol. Chem.* **239,** 3997–4005 (1964).
3. Clark, D. S., Submerged citric acid fermentation of sugar beet molasses: effect of ferrocyanide control, *Ind. Engng. Chem. ind. (int.) Edn.* **1,** 59–62 (1962).
4. Clark, D. S. and Leutz, C. P., Submerged citric acid fermentation of beet molasses in tank-type fermenters, *Biotechnology and Bioengineering* **5,** 193–9 (1963).
5. Ernster, L. and Lee, C., Biological oxidoreductions, *A. Rev. Biochem.* 729–88 (1964).
6. Foote, J. L., Christner, J. E. and Coon, M. J., Biotin- and adenosine-triphosphate-dependent activation of propionyl apocarboxylase, *Biochem. biophys. Acta* **67,** 676–8 (1963).
7. Foster, J. W. and Waksman, S. A., The specific effect of zinc and other heavy metals on growth and fumaric acid production by *Rhizopus*, *J. Bact.* **37,** 599–617 (1939).
8. Huang, H. T., Microbial production of amino acids, *Prog. ind. Microbiol.* **5,** 55–92 (1964).
9. Keech, D. B. and Utter, M. F., Pyruvate carboxylase: II. Properties, *J. biol. Chem.* **238,** 2609–14 (1963).
10. Kinoshita, S., The production of amino acids by fermentation processes, *A. appl. Microbiol.* **1,** 201–14 (1959).
11. Lee, W. H. and Good, R. C., *Amino Acid Synthesis*, U.S.A. patent, 3087863 (1963).
12. Moyed, H. S. and Umbarger, H. E., Regulation of biosynthetic pathways, *Physiol. Rev.* **42,** 444–66 (1962).
13. Nakayama, K., Kitada, S., Sato, Z. and Kinoshita, S., Induction of nutritional mutants of glutamic acid bacteria and their amino-acid accumulation, *J. gen. appl. Microbiol. Tokyo* **7,** 41–51 (1961).
14. Pfeifer, V. F., Vojnovich, C. and Heger, E. N., Itaconic acid by fermentation with *Aspergillus terreus, Ind. Engng. Chem. ind. (int.) Edn.* **44,** 2975–80 (1952).

15. TSUNODA, T., SHIO, I. and MITSUGI, K., Bacterial formation of L-glutamic acid from acetic acid in the growing culture medium: I. Cultural conditions, *J. gen. appl. Microbiol. Tokyo* **7**, 18–29 (1961).
16. TSUNODA, T., SHIO, I. and MITSUGI, K., Bacterial formation of L-glutamic acid from acetic acid in the growing culture medium: II. Growth and L-glutamate accumulation in a chemically defined medium and some other fermentation products, *J. gen. appl. Microbiol. Tokyo* **7**, 30–40 (1961).
17. UTTER, M. F. and KEECH, D. B., Pyruvate carboxylase: I. Nature of the reaction. *J. biol. Chem.* **238**, 2603–8 (1963).
18. WAKSMAN, S. A., *Process for the Production of Fumaric Acid*, U.S.A. patent 2326986 (1943).
19. *A Method of Producing Citric Acid by Fermentation*, British Patent, 841420 (1960).
20. *Preparation of Glutamic Acid*, British Patent, 923200 (1963).

CHAPTER 13

Production of Penicillin Antibiotics

IT IS remarkable, and indeed fortunate for the subsequent progress of human medicine, that the late Sir Alexander Fleming's observation of antibiosis at St. Mary's Hospital, London, in 1928, depended on production by a contaminating culture of an antibiotic having the high antibacterial activity and low mammalian toxicity of penicillin.

1. The general structure of the penicillin antibiotics

The penicillin antibiotics are a closely related group of substances, in which the individual members are distinguished by amide linkage of diverse side-chains to a common fused β-lactam-thiazolidine ring system, known as 6-aminopenicillanic acid (Fig. 13.1).

The first crude preparations of the antibiotic were obtained from surface cultures of *Penicillium notatum* grown on simple media of the Czapek–Dox type that produced predominantly penicillin F (Δ^2-pentenyl-penicillin). The discovery in 1941 of the beneficial effect of cornsteep liquor on the yield of antibiotic was soon followed by the realization that a minor constituent, β-phenylethylamine (see Chapter 6), in this complex source of organic nitrogen directed the fermentation towards production of penicillin G (benzylpenicillin), still the most important penicillin of commerce.

Change to the use of mutant strains of *P. chrysogenum* and the advent of the technique of submerged fermentation in 1944–45 laid

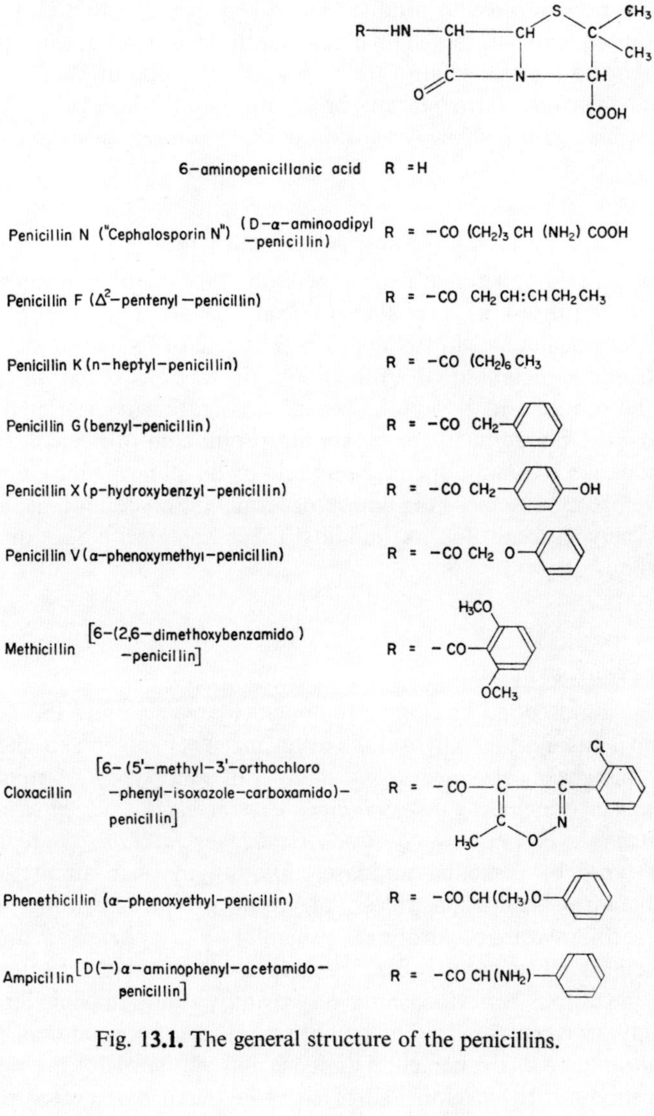

Fig. **13.1.** The general structure of the penicillins.

the foundation of the vast branch of the pharmaceutical industry that is now devoted to producing many medicinal antibiotics.

Subsequently it was shown that addition of appropriate precursors of the side-chain (preferably as an amide or substituted acetic acid) to fermentation broth produced some thirty new penicillins to the virtual exclusion of the "natural" penicillins.

2. 6-Aminopenicillanic acid

Omission of precursor from a medium otherwise able to support the biosynthesis of penicillin G leads to accumulation of 6-aminopenicillanic acid in the broth. The rate of formation of this hydrophilic metabolite is a function of the rate of oxygen transfer to the mould and is optimal when the aeration is restricted to below the requirement for maximum production of penicillin G. Production of 6-aminopenicillanic acid by direct fermentation will therefore require close regulation of fermentation conditions, and this may account for its inability to compete with less direct methods of production.

2.1. *Penicillin amidase*

The amide bond linking 6-aminopenicillanic acid and the side-chain is susceptible to hydrolysis by an enzyme, penicillin amidase; this has been obtained from many different types of micro-organism, including *Penicillium chrysogenum*, *Streptomyces lavendulae*, *Escherichia coli* and *Alcaligenes faecalis*. There are two types of penicillin amidase, differing in their affinity for particular penicillin substrates: the enzyme produced by fungi and by actinomycetes hydrolyses penicillin V (α-phenoxymethyl-penicillin) more rapidly than benzylpenicillin, but the bacterial enzyme utilizes benzylpenicillin very readily, although penicillin V is only attacked slowly. The equilibrium of the reaction depends on the source of the penicillin amidase and on the pH of the reaction mixture. In general, hydrolysis is favoured by elevated pH.

Source of enzyme	Optimum pH		Optimum temperature
	Hydrolysis	Synthesis	
Escherichia coli	8·0	5·5	40°C
Streptomyces lavendulae	9·0	5·5	50°C

2.2. *Partly synthetic penicillins*

Commercially produced by enzymic hydrolysis of benzylpenicillin with the bacterial enzyme, 6-aminopenicillanic acid is used for the chemical synthesis of various novel penicillins: about six of these have some properties that afford a therapeutic advantage over benzylpenicillin and some of the more important of them are listed below.

New penicillin	Therapeutic advantage
Methicillin: 6-(2,6-dimethoxy-benzamido)-penicillin	Resistant to penicillinase
Cloxacillin: 6-(5′-methyl-3′ orthochlorophenyl-isoxazole-carboxamido)-penicillin	Resistant to penicillinase and acid-resistant
Phenethicillin: α-phenoxyethyl-penicillin	Acid-resistant and suitable for oral administration
Penicillin V: α-Phenoxymethyl-penicillin	Acid-resistant and suitable for oral administration
Ampicillin: D(—)α-aminophenyl-acetamido-penicillin	Increased activity against gram-negative bacilli

The partly synthetic penicillins are made commercially by adaptations of one basic method. The 6-aminopenicillanic acid is dissolved in 3% w/v sodium bicarbonate solution containing up to 3% v/v acetone: the appropriate acyl chloride (e.g. α-phenoxy-propionyl chloride for phenethicillin) is added slowly with continuous stirring until excess of the side-chain forming reagent is present. The acylation reaction is completed within 20 min, and the surplus acyl chloride is removed by solvent extraction. The reaction mixture is acidified, and the penicillin present is extracted into solvent and subsequently purified.

3. Biosynthesis of penicillin antibiotics

The biosynthesis of the penicillin antibiotics may be conveniently considered in relation to the side-chain precursors and to the biosynthesis of the fused β-lactam-thiazolidine ring nucleus of 6-aminopenicillanic acid.

3.1. *Precursors of the side-chain*

Side-chain precursors are relatively unspecific, but all of them need to have a terminal -CH_2COOH group, or its metabolic equivalent, to be devoid of toxic action on the fungus and to resist oxidation in the broth. Sometimes these objectives may be conveniently attained by supplying the precursor intermittently in small amounts (e.g. phenylacetic acid in production of benzylpenicillin) or by providing the precursor in a non-toxic masked form that is continuously but slowly metabolized to the active compound (e.g. oxidation of β-phenylethylamine to give phenylacetic acid for biosynthesis of benzylpenicillin, or hydrolysis of *N*-(2-hydroxyethyl)-phenoxyacetamide to give the phenoxyacetic acid precursor needed for biosynthesis of α-phenoxymethylpenicillin).

In practice, the number of different types of penicillin that can be produced by a direct fermentation process is limited by adverse side-effects produced by the precursor. Apart from the toxic effect on the fungus, already considered, the precursor may have adverse effects on the overall biosynthesis by creating an unfavourable pH or an unfavourable redox potential.

3.2. *Biosynthesis of the ring nucleus of* 6-*aminopenicillanic acid*

It may be concluded from the results of numerous experiments with radioactive isotopes, metabolic inhibitors and competitive metabolites that L-cystine can serve as an intermediate in biosynthesis of a substantial portion of the nucleus and that the partial structure illustrated below is incorporated intact:

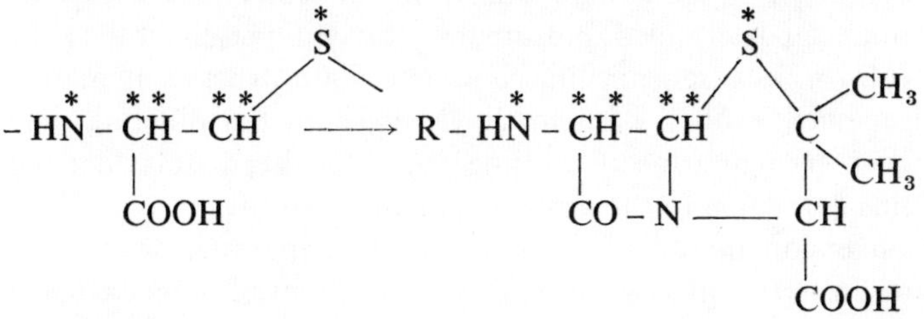

Similarly, L-valine has been found to be a building-block for the remainder of the nucleus, the carbon chain and nitrogen atom of the amino acid being incorporated intact:

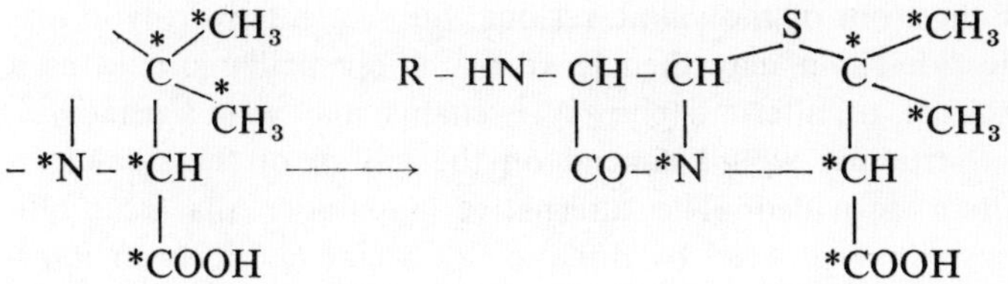

3.3. *The linkage of the component intermediates*

The order and the metabolic forms in which the side-chain, the cystine-derived intermediate, and the valine-derived intermediate are linked to form a penicillin is not known. Some clue to the nature of this biosynthetic process was provided when it was shown that the dipeptide, L-cystinyl-L-valine (but not L-cystinyl-D-valine) was used by mycelium of *Penicillium chrysogenum* for biosynthesis of penicillins by a pathway that did not involve breakdown to the corresponding free amino acids.

The possibility that the biosynthesis of penicillins may progress from the dipeptide to a tripeptide is supported by the isolation from *P. chrysogenum* broths of a tripeptide, δ-(α-aminoadipyl)-cyst(e)inyl-valine and of the corresponding end-product, isopenicillin N (L-α-aminoadipyl-penicillin). Unfortunately, the configuration of the α-aminoadipyl component of the tripeptide has not been determined: presumably, this could be either the L- or the D-form because penicillin N, an antibiotic structurally related to isopenicillin N (L-α-aminoadipyl-penicillin) but differing only in the D-configuration of its side-chain, has been isolated from moulds belonging to the genus *Cephalosporium*.

Conversion of the postulated tripeptide intermediate to isopenicillin N requires two successive cyclizations. The biochemical mechanisms concerned are not known but any satisfactory explanation must account for retention of the α-hydrogen atom and one of the β-hydrogen atoms of the cystine-derived moeity, for lack of migration of any of its hydrogen atoms to the valine moiety, and for inversion from the L- to the D-form, of the α-carbon atom of the valine residue. Subsequent hydrolytic loss of the side-chain may then leave 6-aminopenicillanic acid as a by-product, or, alternatively, acyl exchange may yield a penicillin, whose structure would depend on the nature of the side-chain precursor provided. The alternative possibility that the biosynthesis may proceed by donation of a side-chain to an intermediate 6-aminopenicillanic acid seems less likely because the ability of *Penicillium chrysogenum* to convert 6-aminopenicillanic acid to penicillins in the presence of side-chain precursors has not yet been demonstrated.

4. The course of the fermentation

Mutant strains of *P. chrysogenum* with cornsteep liquor as nitrogen source are normally employed for producing the penicillin antibiotics, of which benzylpenicillin is the major product of direct fermentation.

Crude sucrose (2–3% w/v) is usually the carbohydrate source in the inoculum development medium, but lactose (or an intermittent feed of a more rapidly utilized carbohydrate) is substituted in the production medium, which must also be provided with the requisite precursor.

About 1·5% v/v of a suspension of conidia, produced by shake-flask culture in a medium containing 2·5% w/v calcium chloride as a stimulant to sporulation, is used to inoculate the first inoculum development stage. Transfers (10% v/v) to other inoculum development stages are made as necessary towards the end of each logarithmic phase, until an adequate volume (10% v/v) is available for inoculation of the production medium.

A typical production medium would approximate to the composition: lactose, 3·5% w/v; glucose, 1·0%; cornsteep liquor solids, 3·5%; calcium carbonate, 1·0%; potassium dihydrogen phosphate, 0·4%; and vegetable or animal oil, 0·25% v/v; with post-sterilization pH 5·5–6·0.

The calcium (supplied as some form of calcium carbonate) and the inorganic phosphate of the production medium function mainly as buffers, but their actions are complex. The resultant pH of the broth will be the summation of effects produced by the cations ($Na^{\oplus}$, $K^{\oplus}$, $Ca^{\oplus\oplus}$, $Mg^{\oplus\oplus}$) and anions (phosphate ions), and the extent to which these ions are bound by proteins and solid particles; the degree to which calcium and magnesium ions are precipitated as insoluble phosphates; the pH-dependent decomposition of calcium carbonate; the concentration of ammonium ions; the concentration of precursor acid; and the amounts of metabolic products, including organic acids and penicillins, produced by the mould.

The batch fermentation in the production medium consists of a growth phase followed by a phase of maturation in which most of the penicillin is produced. Rapid proliferation of mycelium occurs in the growth phase, and ammonia is released into the medium as the non-ammonia nitrogen is attacked: the pH rises sharply from an initial value of 4·8 to about 7·5. The maturation phase begins after about 48–60 hr, when the period of rapid growth comes to an end. The maturation phase is characterized by more rapid utilization of lactose and by the consumption of ammonia nitrogen. Lactate is not normally consumed until the sugars have been depleted, so that a low redox potential in the broth is to be expected. This is the main period of penicillin production, and it is important that the pH should approximate to 7·0 throughout and that the temperature should be carefully regulated (25 ±0·5°C).

The conditions (high redox potential and availability of ammonium ion) observed during the maturation phase will clearly favour the reductive amination by glutamic acid dehydrogenase of α-ketoglutarate to L-glutamic acid (see Chapter 12), thereby tending to limit operation of the citrate cycle because of a deficiency of C_4 intermediates. In these circumstances, oxidation by the citrate cycle of acetyl-coenzyme A would be diminished, with consequent diversion of pyruvate to α-acetolactate, the first of the intermediates needed for biosynthesis of L-valine.

Oil supplied to the fermentation broth may be expected to have several beneficial effects. The β-oxidation of the fatty acids, with concomitant reduction of the FAD and NAD coenzymes, may help to maintain adequate reducing potential in the system. Metabolism of the fatty acids to acetyl-coenzyme A may saturate the demands for substrate of the citrate cycle, thereby sparing pyruvate from oxidative decarboxylation and permitting its diversion to α-acetolactate (see Chapter 10). Finally, the action of oil may alter the morphology of the mycelium forming the wide mycelium and the spherical bodies described by Duckworth and Harris; and these may help to prolong the viability of the micro-organism in the broth, especially when provision is made to maintain simultaneously the nitrogen content of the medium above a minimum value of about 1 mg/ml.

Suggested literature for further reading

1. ABRAHAM, E. P. and NEWTON, G. G. F., New penicillins, cephalosporin C, and penicillinase, *Endeavour* **20,** 92–100 (1961).
2. ARNSTEIN, H. R. V. and GRANT, P. T., The biosynthesis of penicillin. I. The incorporation of some amino acids into penicillin, *Biochem. J.* **57,** 353–9 (1954).
3. ARNSTEIN, H. R. V. and GRANT, P. T., The biosynthesis of penicillin. 2. The incorporation of cystine into penicillin, *Biochem J.* **57,** 360–8 (1954).

4. Arnstein, H. R. V. and Clubb, M. E., The biosynthesis of penicillin. 5. Comparison of valine and hydroxyvaline as penicillin precursors, *Biochem. J.* **65,** 618–27 (1957).
5. Arnstein, H. R. V. and Crawhall, J. C., The biosynthesis of penicillin. 6. A study of the mechanism of the formation of the thiazolidine-β-lactam rings, using tritium-labelled cystine, *Biochem. J.* **67,** 180–7 (1957).
6. Arnstein, H. R. V. and Morris, D., The utilization of L-cystinyl-L-valine for penicillin biosynthesis, *Biochem. J.* **76,** 323–7 (1960).
7. Arnstein, H. R. V., Artman, M., Morris, D. and Toms, E. J., Sulphur-containing amino acids and peptides in the mycelium of *Penicillium chrysogenum, Biochem. J.* **76,** 353–7 (1960).
8. Arnstein, H. R. V. and Morris, D., The structure of a peptide, containing α-aminoadipic acid, cystine and valine present in the mycelium of *Penicillium chrysogenum, Biochem. J.* **76,** 357–61 (1960).
9. Batchelor, F. R., Chain, E. B., Richards, M. and Rolinson, G. N., 6-aminopenicillanic acid. VI. Formation of 6-aminopenicillanic acid from penicillin by enzymic hydrolysis, *London Proc. R. Soc.* B, **154** (957), 522–31 (1961).
10. Behrens, O. T., Corse, J., Edwards, J. P., Garrison, L., Jones, R. G., Soper, Q. F., Van Abeele, R. F. and Whitehead, C. W., Biosynthesis of penicillins. IV. New crystalline biosynthetic penicillins, *J. biol. Chem.* **175,** 793–809 (1948).
11. Birch, A. J. and Smith, H., Oxidative formation of biologically active compounds from peptides (in Ciba Symposium: *Amino Acids and Peptides with Antimetabolic Activity*, pp. 247–60), Churchill, London, 1958.
12. Demain, A. L., The mechanism of penicillin biosynthesis, *Adv. appl. Microbiol.* **1,** 23–47 (1959).
13. Demain, A. L., Inhibition of penicillin formation by amino acid analogs, *Archs. Biochem. Biophys.* **64,** 74–9 (1956).
14. Duckworth, R. B. and Harris, G. C. M., The morphology of *Penicillium chrysogenum* in submerged fermentations, *Trans. Brit. Mycol. Soc.* **32,** 224–35 (1949–50).
15. Flynn, E. H., McCormick, M. H., Stamper, M. C., De Valera, H. and Godzeski, C. W., A new natural penicillin from *Penicillium chrysogenum, J. Am. chem. Soc.* **84,** 4594–5 (1962).
16. Freany, T. E., *Process for Producion of Penicillin*, U.S.A. Patent, 2830934 (1955).
17. Hockenhull, D. J. D., The influence of medium constituents on the biosynthesis of penicillin, *Prog. ind. Microbiol.* **1,** 1–27 (1959).
18. Leonard, N. J. and Wilson, Jr. G. E., The synthesis and oxidative re-arrangement of some 1,4-thiazepines related to the penicillins, *J. Am. chem. Soc.* **86,** 5307–16 (1964).
19. Moss, M. O. and Cole, M., Reactions of 6-aminopenicillanic acid with carbohydrates and related substances, *Biochem. J.* **92,** 643–8 (1964).
20. Perron, G. G., Minor, W. F., Holdrege, C. T., Gottstein, W. J., Godfrey, J. C., Crast, L. B., Babel, R. B. and Cheney, L. C., Derivatives of 6-aminopenicillanic acid. I. Partially synthetic penicillins prepared from α-aryloxyalkanoic acids, *J. Am. chem. Soc.* **82,** 3934–8 (1960).

21. RADHAKRISHNAN, A. N., and SNELL, E. E., Biosynthesis of valine and isoleucine. II. Formation of α-acetolactate and α-aceto-α-hydroxybutyrate in *Neurospora crassa* and *Escherichia coli*, *J. biol. Chem.* **235**, 2316–21 (1960).

22. RADHAKRISHNAN, A. N., WAGNER, R. P. and SNELL, E. E., Biosynthesis of valine and isoleucine. III. α-keto-β-hydroxy acid reductase and α-hydroxy-β-keto acid reductoisomerase, *J. biol. Chem.* **235**, 2322–30 (1960).

23. RICHARDS, G. N. R. M. and BATCHELOR, F. R., *Enzymic Production of 6-Aminopenicillanic Acid*, British Patent, 892144 (1962).

24. ROLINSON, G. N., BATCHELOR, F. R., BUTTERWORTH, D., CAMERON-WOOD, J., COLE, M., EUSTACE, G. C., HART, M. V., RICHARDS, M. and CHAIN, E. B., Formation of 6-aminopenicillanic acid from penicillin by enzymatic hydrolysis, *Nature. Lond.* **187**, 236–7 (1960).

25. SINGH, K. and JOHNSON, M. J., Evaluation of precursors for penicillin G, *J. Bact.* **56**, 339–55 (1948).

26. STEVENS, C. M., VOHRA, P. and DELONG, C. W., Utilization of valine in the biosynthesis of penicillins, *J. biol. Chem.* **211**, 297–300 (1954).

27. STEVENS, C. M., INAMINE, E. and DELONG, C. W., The rates of incorporation of L-cystine and D- and L-valine in penicillin biosynthesis, *J. biol. Chem.* **219**, 405–9 (1956).

28. STEVENS, C. M. and DELONG, C. W., Valine metabolism and penicillin biosynthesis, *J. biol. Chem.* **230**, 991–9 (1958).

29. SYLVESTER, J. C. and COGHILL, R. D., The penicillin fermentation (Chapter 7 in *Industrial Fermentations*), 219–63, Chemical Publishing Co., New York, 1954.

30. WOLFF, E. C. and ARNSTEIN, H. R. V., The metabolism of 6-aminopenicillanic acid and related compounds by *Penicillium chrysogenum* and its possible significance for penicillin biosynthesis, *Biochem. J.* **76**, 375–81 (1960).

CHAPTER 14

Biosynthesis from Acetate in Relation to Production of Griseofulvin and the Tetracycline Antibiotics

1. The use of two-carbon units in biosynthesis

Two carbon units are used as building units in two important biosynthetic pathways, and these will be considered in turn.

1.1. *Biosynthesis of isoprenoids*

Mevalonic acid, formed by condensation of three molecules of acetyl-coenzyme A, is a key intermediate in biosynthesis of isopentenyl pyrophosphate. This "active isoprene unit" is used as the C_5 building unit for biosynthesis of the long-chain intermediates that form steroids, carotenoids and terpenes and for introduction of a C_5 unit by *O*- and *C*-acylation:

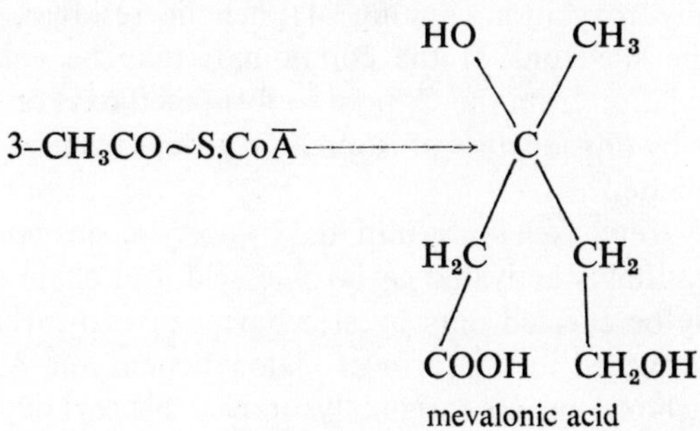

mevalonic acid

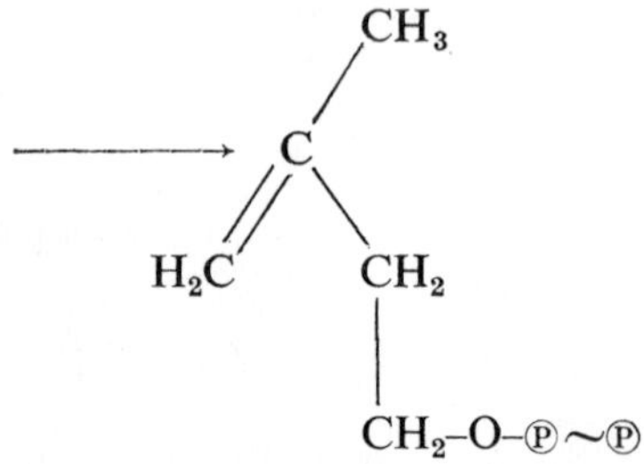

isopentenylpyrophosphate
(active isoprene unit)

1.2. *Linear condensation of two-carbon units into poly-β-keto-methylene intermediates for biosynthesis of aromatic compounds: the acetate theory*

The skeletons of many natural phenolic compounds are considered to be derived, at least in part, by head-to-tail linkage of acetic acid units followed by different types of ring closure (Fig. 14.1). Phloroglucinol derivatives could be formed by reactions of the *C*-acylation type; cyclization by an aldol-type reaction could lead, for example, to orcinol derivatives; and pyrone nuclei could be formed by interaction between carbonyl groups. The theory has proved so reliable that, given the carbon chain, it is frequently possible to predict the location of oxygen substituents, but it must be remembered that the pattern of oxygenation may be masked by combinations of oxidation and reduction, together with aromatic hydroxylation, linking of phenolic residues, and ring fission. The skeletons of the compounds may be enlarged by transmethylation from the C_1 pool to form methoxyl or *C*-methyl groups, or by introduction of isoprenoid groups from isopentenyl-pyrophosphate.

The polyacetyl skeleton is initiated by acetyl-coenzyme A or by any other suitably activated carboxylic acid, but chain extension can usually be effected only by attachment of two-carbon units: these are attached in the form of malonyl-coenzyme A, and the surplus carboxyl group (originally derived by carboxylation of

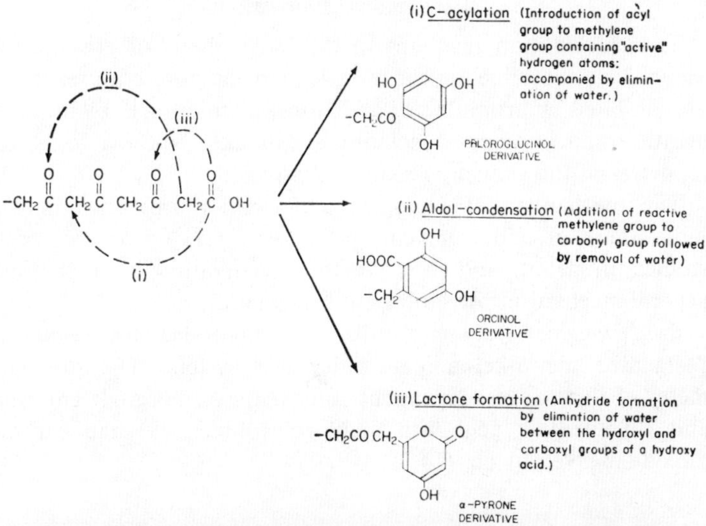

FIG. 14.1. Cyclization of head-to-tail linked two-carbon units (the acetate theory of biosynthesis).

acetyl-coenzyme A) is eliminated again as soon as it has provided the energy needed for chain extension.

Failure to detect intermediate steps between the final carbon skeleton and the activated building-blocks employed in its biosynthesis from two-carbon units has led to the suggestion that a poly-β-ketomethylene chain precursor may be constructed, cyclized, and sometimes modified *in situ* on the enzyme surface.

Acceptance of this hypothesis would imply that the biosynthesis of compounds such as griseofulvin and the tetracycline antibiotics is essentially complete before the primary skeleton of the product is liberated. It should, however, be noted that there is no present evidence to exclude a C_8 open chain or carbocyclic compound as being the common intermediate from which biosynthetic routes to numerous acetate-derived phenolic compounds may depart at some stage before aromatization.

2. Biological chlorination

The chlorination reactions in the biosynthesis of griseofulvin and chlortetracycline appear to be similar, because both reactions are inhibited by bromide or iodide ions, by thiouracil, by reduced aeration and by copper-chelating substances. Nothing has been reported on the enzymic reactions involved.

The mechanism of biological chlorination operative in the biosynthesis of caldariomycin by *Caldariomyces fumago* has been studied in detail, and it is worth consideration as a possible general mechanism for biological chlorination.

Caldariomyces fumago utilizes a multi-enzyme complex, designated the β-ketoadipate chlorinase system. The stoichiometry of this oxygen-dependent conversion of inorganic chloride and β-ketoadipic acid into δ-chlorolevulinic acid and carbon dioxide is given by eq. (14.1).

$$^{\ominus}OOCCH_2CH_2\overset{\overset{\displaystyle O}{\|}}{C}CH_2COO^{\ominus}+Cl^{\ominus}+\tfrac{1}{2}O_2+2H^{\oplus}$$
$$\longrightarrow CO_2+{}^{\ominus}OOCCH_2CH_2\overset{\overset{\displaystyle O}{\|}}{C}CH_2Cl+H_2O \qquad (14.1)$$

The individual enzymes concerned in the β-ketoadipate chlorinase system are glucose oxidase, catalase, chloroperoxidase and a polysaccharide hydrolase operating according to eqs. (14.2) to (14.5):

$$(\text{Glucose})_n+H_2O \xrightarrow[\text{polysaccharide hydrolase}]{} (\text{glucose})_{n-1}+\text{glucose} \qquad (14.2)$$

$$\text{glucose}+O_2+H_2O \xrightarrow[\text{glucose oxidase}]{} \text{gluconate}+H_2O_2+H^{\oplus} \qquad (14.3)$$

$$2H_2O_2 \xrightarrow[\text{catalase}]{} O_2+2H_2O \qquad (14.4)$$

$$H_2O_2+Cl^{\ominus}+{}^{\ominus}OOCCH_2CH_2COCH_2COO^{\ominus}+2H^{\oplus}$$
$$\xrightarrow[\text{chloroperoxidase}]{} CO_2+{}^{\ominus}OOCCH_2CH_2COCH_2Cl+2H_2O \qquad (14.5)$$

It has been suggested that the chlorination reaction (eq. (14.5)) might require oxidation of chloride ion to a chlorinium ion and that the latter, stabilized as an enzyme-$Cl^{\oplus}$ complex, is transferred to the β-ketoadipic acid. An alternative suggestion is the possibility that the energy of the chlorinium ion may be conserved by formation of δ-chlorolevulinic acid and that this compound may function as a biological chlorine donor in a manner analogous to that occurring in the transfer of energy-rich phosphate groups.

3. Biological transmethylation

In the biosynthesis of griseofulvin, *O*-methylation plays an important part and both *C*- and *N*-methylations are necessary for forming the tetracycline antibiotics: all these processes depend on transmethylations from onium compounds functioning as biological methylating agents or methyl donors.

The prime methyl donor in biological transmethylations is (—)-*S*-adenosylmethionine formed by the enzymic reactions indicated below.

$$\text{L-methionine} + \text{ATP} \xrightarrow[\text{Mg}^{\oplus\oplus}]{} (-)\text{-}S\text{-adenosylmethionine} + \text{PP}_i + \text{P}_i$$

(—)-*S*-adenosylmethionine functions as a coenzyme in biological *N*-transmethylation, *O*-transmethylation, *S*-transmethylation and *C*-transmethylation reactions involved in the formation of many important methylated compounds. Some of these products are able to function in several other methyltransferase reactions that do not require *S*-adenosylmethionine: the methyl donors may be betaine (the oxidation product of choline), dimethylthetin or methylmethionine, but the acceptor of the methyl group is always, L-homocysteine, giving L-methionine as the methylated product.

3.1. De novo *biosynthesis of methionine methyl*

A variety of compounds (e.g. serine, formaldehyde) can donate one-carbon units to tetrahydrofolate with formation of methylene-tetrahydrofolate (active formaldehyde):

(1) L-serine + tetrahydrofolate

$$\underset{\text{L-serine hydroxymethyl-transferase}}{\rightleftharpoons} \text{glycine} + N^5N^{10}\text{-methylenetetrahydrofolate}$$

(2) HCOOH + ATP + tetrahydrofolate

$$\underset{\text{formyl tetra-hydrofolate synthetase}}{\rightleftharpoons} N^{10}\text{-formyltetrahydrofolate} + \text{ADP} + \text{P}_i$$

$$\underset{\text{acidification}}{\longrightarrow} N^5N^{10}\text{-methylenetetrahydrofolate}$$

Methylenetetrahydrofolate provides a metabolic pool of C_1 units that can be used for biosynthesis of methionine methyl, but the first step requires formation of the labile methyl group of N^5-methyltetrahydrofolate.

N^5N^{10}-methylenetetrahydrofolate + NADH + $H^{\oplus}$

$$\underset{N^5N^{10}\text{-methylene-tetrahydrofolate reductase}}{\longrightarrow} N^5\text{-methyltetrahydrofolate} + \text{NAD}^{\oplus}$$

The hydrogen atoms attached to the methylene group of N^5N^{10}-methylenetetrahydrofolate provide two of the hydrogen atoms of the newly synthesized methyl group, but the third hydrogen atom is derived from the protons of the solvent.

The methyl group of N^5-methyltetrahydrofolate is transferred to homocysteine and forms methionine in a complex reaction requiring the presence of *S*-adenosylmethionine and methyl-vitamin B_{12}. It has been suggested that the methyl group of N^5-methyltetrahydrofolate may be transferred as a methyl carbonium ion ($^{\oplus}CH_3$) to reduced vitamin B_{12}-protein having two free electrons, with formation of methyl-vitamin B_{12}-protein: *S*-adenosylmethionine is necessary for this methyl transfer. The biosynthesis of methionine could then be completed by transfer of the methyl group of methyl-vitamin B_{12}-protein, as a carbonium ion, to homocysteine, with concomitant regeneration of reduced vitamin B_{12}-protein.

3.2. *Biosynthesis of choline*

All of the three methyl groups of choline $CH_2OH.CH_2\overset{+\oplus}{N}(CH_3)_3$ are derived from methionine, the methyl donor being *S*-adenosylmethionine and the biosynthesis proceeding at the phosphatide level. The transmethylations occur stepwise: phosphatidylaminoethanol (formed via phosphatidyl serine from serine and diacylglycerophosphate) is first methylated to phosphatidylmonomethylaminoethanol; a second transmethylation gives phosphatidyldimethylaminoethanol, and the third lecithin (phosphatidyl choline).

Removal of the two fatty acid residues from lecithin gives glycerophosphocholine, and this compound can then be cleaved cleaved to liberate choline or, on further oxidation, betaine $[COOH.CH_2\overset{\oplus}{N}(CH_3)_2]$.

Betaine is a particularly useful metabolite. One of the methyl groups of betaine can be donated to homocysteine with formation of methionine in the methyltransferase reaction already described: the remaining two methyl groups, first from dimethylglycine and then from monomethylglycine (sarcosine), can be used as an additional source for stepwise transfer of the C_1 units necessary for formation of active formaldehyde (N^5N^{10}-methylenetetrahydrofolate).

4. Biosynthesis of (+)-griseofulvin

Benzophenones occur as intermediates in the later stages of (+)-griseofulvin biosynthesis (Fig. 14.2). Griseophenone C (2:6:4′-trihydroxy-4:2′-dimethoxy-6′-methylbenzophenone) (I) is chlorinated to give griseophenone B (5-chloro-2:6:4′-trihydroxy-

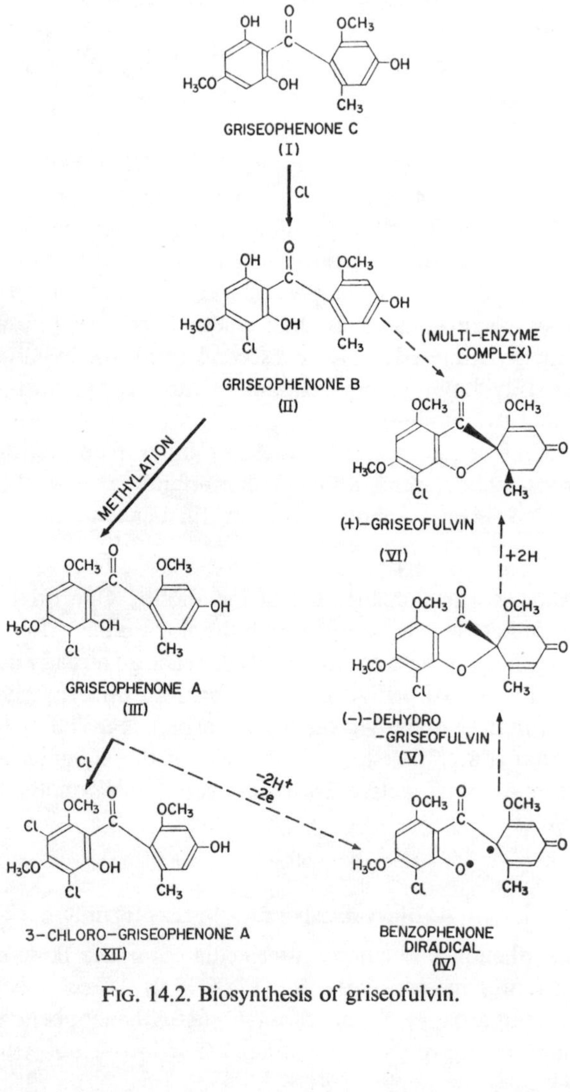

FIG. 14.2. Biosynthesis of griseofulvin.

4:2′-dimethoxy-6′-methylbenzophenone) (II) and then ultimately (+)-griseofulvin (VI).

Another benzophenone, griseophenone A (5-chloro-6:4′-dihydroxy-2:4:2′-trimethoxy-6′-methylbenzophenone) (III), produced in the fermentation by methylation of griseophenone B, has been considered to be a further intermediate in griseofulvin biosynthesis. It has been suggested that oxidation by 1-electron transfer agents could form the benzophenone diradical (IV), which would then be coupled to give the intermediate dehydrogriseofulvin (V) and finally griseofulvin (VI) by enzymic reduction.

This free-radical mechanism for biosynthesis of natural griseofulvin is attractive, but two observations detract cogently from its acceptability at the present time: (i) experiments with radioisotopes have failed to show incorporation of griseophenone A into (+)-griseofulvin, although both griseophenone B and griseophenone C are readily incorporated; (ii) free-radical coupling of griseophenone A by an enzyme obtained from the fungus *Polystictus versicolor* yields (±)-dehydrogriseofulvin instead of the natural isomer, (—)-dehydrogriseofulvin, that occurs in griseofulvin fermentation broths.

In an alternative hypothesis it has been suggested that griseophenone A and (—)-dehydrogriseofulvin may be by-products formed when components of a multi-enzyme complex, responsible for biosynthesis of (+)-griseofulvin from griseophenone B, fail to act in unison. This scheme suggests that a concerted reaction, involving more or less simultaneous oxidation reduction and methylation, begins with adsorption of griseophenone B anion to three specific centres in the multi-enzyme complex: the adsorbed anion could assume the desired configuration for subsequent asymmetric biosynthesis of the coumaranone ring through a carbonium ion intermediate. We should note that continued operation of this multi-enzyme complex would depend on maintenance of a suitable pH, continuous provision of ATP, $NADPH_2$ and oxygen and the regeneration of N^5-methyltetrahydrofolate from the mole of dihydrofolate formed in each cycle.

5. Production of griseofulvin

5.1. *Griseofulvin-producing moulds*

Griseofulvin is firmly established as a non-toxic systemic antifungal antibiotic for the treatment and control of fungal diseases of the skin, hair and nails by oral therapy. The antibiotic is a metabolic product of fungal species belonging to two major divisions of the genus *Penicillium:* Asymmetrica Fasciculata and Asymmetrica Divaricata. The only apparent exception is *P. brefeldianum* Dodge, but some strains of this species are also considered to have a close affinity with the Carpenteles series of the Asymmetrica Divaricata section.

Most of the strains used for commercial production of griseofulvin are mutants of *P. patulum* Bainier-Thom. Spore suspensions can be produced in submerged culture if the medium is well aerated and if the total available nitrogen concentration is limited to between 0·05 and 0·1 % w/v. It is notable that there is a marked interaction between the concentration of available nitrogen and the aeration of the medium: the lower the aeration, the lower the concentration of nitrogen needed to induce sporulation. Sporulation of a griseofulvin-producing strain of *P. griseofulvum* may also be rapidly induced in submerged culture by transferring mycelium to a medium containing a high concentration of glucose but without assimilable nitrogen. The most powerful stimulus to sporulation occurs when mycelium emerges from submerged nito aerial conditions: it has been suggested that sporulation under these conditions may be associated with some physical change at the cell surface caused by abrupt formation of an air–water interface.

Sporulation, although desirable for preparation of inoculum, imposes limitations on the griseofulvin fermentation. Griseofulvin production in pilot-plant or production-type fermenters is greatly depressed when the available nitrogen in the medium is limited to such an extent that profuse sporulation occurs. The situation is much worse in mechanically agitated and aerated laboratory fermenters: mycelium is impelled from the broth to adhere on the glass walls of the fermenters where spores are produced, presumably under the stimulus of abrupt formation of an air–water interface coincident with removal of mycelium from the nitrogen of the medium. For this reason submerged fermentation of griseofulvin, although suitable for shake-flasks and plant-type fermenters, is unsatisfactory in mechanically agitated and aerated 5 l. laboratory fermenters.

5.2. *The griseofulvin fermentation*

Factors of prime importance in the submerged fermentation of griseofulvin are concentration of available nitrogen; aeration; concentrations of phosphate and calcium carbonate; pH; and nature of the carbohydrate, its concentration and its rate of supply.

Optimum yield is obtained when the initial concentration of available nitrogen in the medium is between 0·15 and 0·25% w/v, depending on the degree of aeration and the nature of the carbohydrate. Lactose (7·5–11% w/v) is preferred for batch fermentations, because excess sugar, especially when rapidly metabolized, depresses titre by diversion of biosynthesis to other products. Starch hydrolysate or glucose may, however, be conveniently substituted when provision can be made for limitation of the carbohydrate concentration in the medium by intermittent or continuous feed. Good aeration is essential for griseofulvin production, and this requirement is particularly important when the available nitrogen concentration is relatively high or when the rate of supply of readily available carbohydrate is excessive.

The pH of the fermentation broth depends on the nature of the medium, the rate of aeration and the type and size of the culture vessel used; it is particularly sensitive to changes in the carbohydrate concentration. The pH in griseofulvin fermentation broth is usually controlled by serial additions of readily available carbohydrate as necessary: consumption of carbohydrate by the mould causes the pH to rise, but the effect can be quickly reversed by addition of more carbohydrate. Rates of supply of carbohydrate are so adjusted that the pH of the medium conforms to an experimentally determined "ideal" pH curve, characteristically exhibiting a rapid rise from an initial pH 5·5 to pH 6·8 at the 24th hour, with then a relatively constant pH 6·8–7·2 until the end of the fermentation period.

A typical fermentation medium of pH 5·5 before sterilization (15 min at 121°C) would contain 0·4% w/v limestone, 0·4% KH_2PO_4, 0·15% KCl, 0·005% of a proprietary anti-foam agent and cornsteep liquor to give 0·175% nitrogen. The medium, maintained at 25°C, is inoculated with 10% v/v of vegetative mycelium, and the fermentation period is usually terminated within 14 days.

A 400 gal batch of medium would need an air-flow of about 25 ft^3/min for the first 5 hr, 50 ft^3/min from the fifth to the tenth hour and then about 80 ft^3/min until the end of the fermentation period.

Beginning with no carbohydrate in the medium except that supplied by the cornsteep liquor, glucose syrup (50% w/v aqueous solution of hydrolysed starch) is added from about the ninth hour. Typically, 400 gal of medium might be expected to receive glucose syrup from the ninth hour at the rate of 0·75 l./hr (supplied as 7 portions of 0·1–0·3 l. given at intervals of 2, 4 or 8 min). The pH should continue to rise rapidly, reaching pH 6·8 at 20–24 hr. The rate of addition of glucose syrup is then increased to about 4·0 l./hr, minor adjustments being made to maintain the pH of the medium in conformity with the substantially horizontal portion of the "ideal" pH curve until the end of the fermentation period. Broth yields of griseofulvin up to 5·8 g/l. have been reported.

Improved yields of griseofulvin are obtained when 90–95% of the assimilable nitrogen of the medium, used within the concentration range already defined, is provided from mixed sources of complex organic nitrogen (e.g. cornsteep liquor with distiller's solubles), and when the remaining 5–10% is supplied as ammonia, as an ammonium salt or as urea. A suitable medium of this type capable of giving about 12 g/l. griseofulvin in 13 days contains cornsteep liquor, distiller's solubles and ammonium sulphate in amounts giving 0·075%, 0·075% and 0·025% nitrogen respectively.

The methylation reactions necessary for biosynthesis of griseophenone C can, if needed, be greatly accelerated by including in the medium a methyl donor such as choline hydrochloride (about 0·1–0·2% w/v).

5.3. *By-products of the griseofulvin fermentation* (Fig. 14.3)

Dechlorogriseofulvin (VII) is usually present in the broth to a small extent, and hydrolysis of this compound may account for the occurrence of some dechlorogriseofulvic acid (VIII).

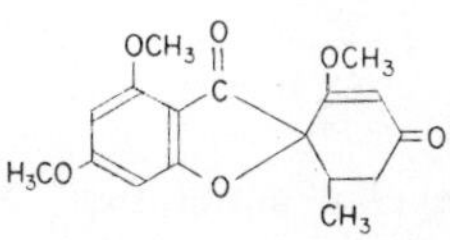

DECHLOROGRISEOFULVIN (VII)

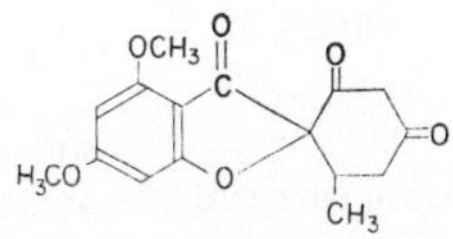

DECHLOROGRISEOFULVIC ACID (VIII)

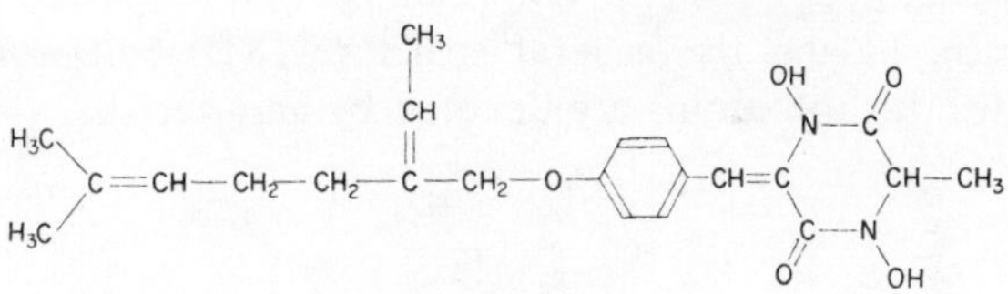

MYCELIANAMIDE (IX)

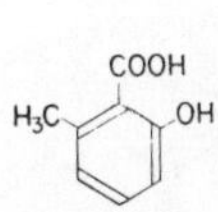

6-METHYLSALICYLIC ACID (X)

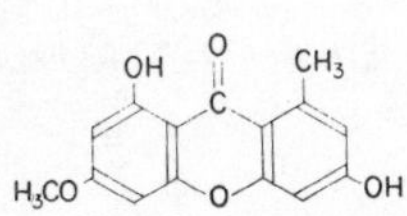

GRISEOXANTHONE C (XI)

Fig. 14.3. By-products of the griseofulvin fermentation process.

Mycelianamide (IX), a terpenoid containing nitrogen, ordinarily accompanies griseofulvin in the broth, but the relative amounts formed vary with the fermentation medium and with the strain employed.

The biosynthesis of griseofulvin is prone to be diverted under unfavourable conditions, permitting the formation of other unwanted by-products. Excessively rapid sugar consumption in the early period of fermentation may give some 6-methylsalicylic acid (X), and any deficiency in the chlorination reaction will be manifested by accumulation of griseophenone C with subsequent acceleration of side-reactions forming griseoxanthone C (XI). On the other hand, failure in the later stages of griseofulvin biosynthesis may account for the presence in broth of abnormally high concentrations of (—)-dehydrogriseofulvin (V) and griseophenone A (III), as well as for the occurrence of a small quantity of chlorogriseophenone A (XII).

6. Structures of the tetracycline antibiotics

The tetracycline antibiotics comprise a closely related group of broad-spectrum antibacterial compounds of immense commercial importance, having the general structure (XIII), in which the asymmetric carbon atoms are denoted by an asterisk:

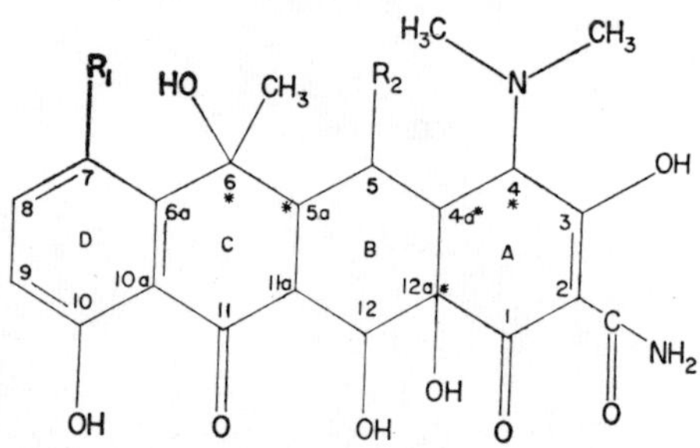

(XIII)

Tetracycline (XIII; $R_1 = R_2 = H$)
Oxytetracycline (Terramycin) (XIII; $R_1 = H$, $R_2 = OH$)
Chlortetracycline (Aureomycin) (XIII; $R_1 = Cl$, $R_2 = H$)
Bromtetracycline (XIII; $R_1 = Br$, $R_2 = H$)

The C-6 methyl group is absent in two important demethyltetracyclines, 6-demethyltetracycline and 7-chloro-6-demethyltetracycline: the C-7 chlorine atom is absent in 7-dechlorotetracycline.

The carboxamide group at C-2 is present in all the commercially important tetracycline antibiotics, but oxytetracycline X (2-acetyl-2-decarboxamido-oxytetracycline), having an acetyl group replacement at position 2, has been isolated from fermentation broth.

7. Biosynthesis of the tetracycline antibiotics

It was suggested on the basis of early experiments with radioactive isotopes that the ring structure of oxytetracycline (XIII, R_1=H, R_2=OH) is derived mainly from acetate units, with glutamate forming the carboxamide group and part of ring A (carbon atoms 2, 3, 4 and 4a). Later work indicates that the whole tetracycline nucleus is derived by condensation of C_2 units according to expectation from the acetate theory, but that the primer for biosynthesis of the polyketide chain is probably the amide of malonyl-coenzyme A (providing the carboxamide group and carbon atoms 1 and 2). This method of biosynthesis also affords an explanation for the co-existence in the broth of 2-acetyl-2-decarboxamido-oxytetracycline: here the primer would be expected to be acetyl-coenzyme A instead of malonamyl-coenzyme A.

The nitrogen of the dimethylamino residue at position 4 in ring A is introduced by transamination. The C-6 methyl group and both of the *N*-methyl groups are formed by *C*-methylation and by *N*-methylation, respectively, from the biological C_1-donor system already described. The chlorine atom at position 7 in the chlortetracycline molecule is introduced during the later stages of biosynthesis: the intermediate concerned has not been identified, but the possibility that tetracycline may be chlorinated can be excluded.

The nature of the ring systems that precede formation of the tetracycline nucleus has not been completely elucidated. A fully aromatic intermediate appears to be concerned, because certain naphthacenic compounds (XIV) can be converted microbiologically to tetraçyclines:

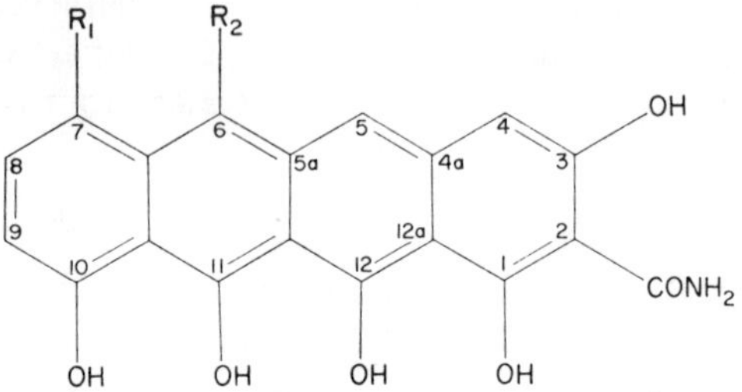

(XIV)

R_1 = H or Cl
R_2 = H or CH_3

Chlorinated tetracyclines are formed either from a chlorinated naphthacenic intermediate or from an unchlorinated naphthacenic intermediate supplied to a chlorinating strain. Unchlorinated tetracyclines arise only when non-chlorinated intermediates are provided for non-chlorinating strains. Conversely, 6-methylated tetracyclines cannot be formed from unmethylated naphthacenic intermediates even when the strain employed is not blocked for the C-6 methylation step in tetracycline biosynthesis. Therefore unmethylated naphthacenic intermediates afford unmethylated tetracyclines, and the corresponding methylated intermediates are converted to methylated tetracyclines. From these results it was concluded that introduction of the C-6 methyl group in tetracycline biosynthesis must occur before formation of the naphthacenic intermediate, but that the C-7 chloro group must be introduced at a subsequent stage.

The steps in tetracycline biosynthesis between the naphthacenic intermediate and the tetracycline nucleus have not yet been completely determined; the process must be complex because all five of the asymmetric centres are generated at this time.

At some stage, the aromaticity of the naphthacenic intermediate must be broken, but the nature of the intermediates concerned has not been reported.

The last two steps in the biosynthesis of tetracyclines appear to be the oxidative hydroxylation at C-6 of 5a, 6-anhydrotetracyclines (XV) to give dehydrotetracyclines (XVI), the immediate

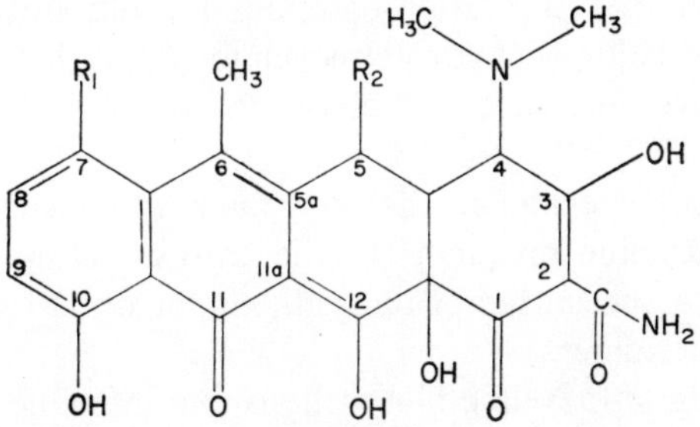

(XV)

Chlortetracycline intermediate (XV; R_1 = Cl, R_2 = H)
Presumed tetracycline intermediate (XV; R_1 = R_2 = H)
Presumed oxytetracycline intermediate (XV; R_1 = H, R_2 = OH)

precursors that can undergo microbiological reduction to the tetracyclines:

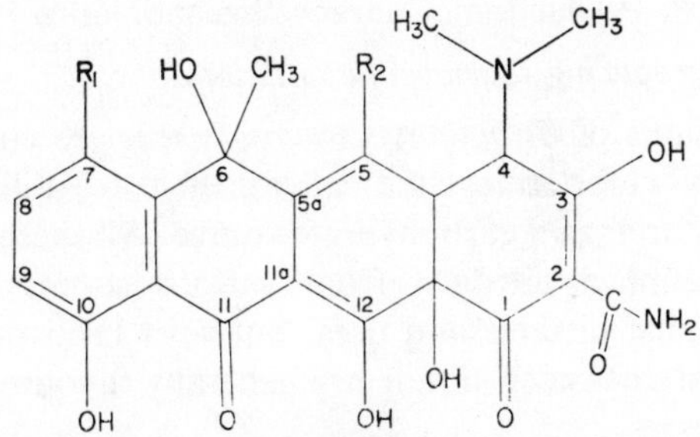

(XVI)

Chlortetracycline precursor (XVI; R_1 = Cl, R_2 = H)
Presumed tetracycline precursor (XVI; R_1 = R_2 = H)
Presumed oxytetracycline precursor (XVI; R_1 = H, R_2 = OH).

The original work suggested that the penultimate step may yield a 5a,11a-dehydrotetracycline, but more recent work, which provides a chemical analogy for the biosynthesis, suggests that the double bond is 5,5a as in XVI, rather than 5a,11a. The chemical synthesis showed that 5a, 6-anhydro-7-chloro-tetracycline(XV, $R_1 = Cl$, $R_2 = H$) undergoes photo-oxidation to the 6-hydroperoxide and that the latter compound could be catalytically reduced to 7-chloro-dehydrotetracycline (XVI; $R_1 = Cl$, $R_2 = H$), the known precursor of chlortetracycline (XIII, $R_1 = Cl$, $R_2 = H$).

There is some evidence that the final hydrogenation of the dehydrotetracycline precursor to the corresponding tetracycline may require a special factor (cosynthetic factor 1) related to the flavines or pteridines.

The stage in tetracycline biosynthesis that introduces the nitrogen at position 4 has not been determined, but present evidence suggests that this does not occur until the naphthacenic intermediate has been completely formed. The complete dimethylamino group in the proper configuration is essential in the 5a,6-anhydrotetracyclines that serve as substrates for the last two stages of the biosynthesis.

8. Production of tetracycline antibiotics

8.1. *Moulds producing tetracycline antibiotics*

Mutant strains of *Streptomyces aureofaciens* are employed for production of chlortetracycline: all are able to utilize glucose, sucrose and starch as a carbohydrate source. All these strains are able to metabolize nitrogen from various sources, including ammonium salts, nitrates and urea, but soya bean meal, cotton seed meal and cornsteep liquor are generally favoured for commercial processes.

Mutants having the best potential for production of chlortetracycline form colonies that are characteristically non-sporulating and highly pigmented (yellow to orange–yellow).

Strains of *S. aureofaciens* may be distinguished according to their ability to utilize chloride and bromide ions. "Scavenging" strains can use all the chloride ions present in the medium to produce predominantly chlortetracycline with only a little tetracycline. In the absence of chloride these strains are able to use bromide ions, with formation of bromtetracycline instead of chlortetracycline. When both chloride and bromide are supplied in the medium the bromide ions are no longer used for biosynthesis of bromtetracycline: instead, the bromide ions now inhibit the biosynthesis of chlortetracycline, with the result that the yield of tetracycline in increased.

The "non-scavenging" strains are unable to produce bromtetracycline, so that these strains tend to produce moderate amounts of tetracycline whenever the biosynthesis of chlortetracycline is inhibited.

Production of the demethyl tetracyclines depends on the employment of mutant strains having the required specific genetic block in biosynthesis of chlortetracycline or of tetracycline.

Commercial production of oxytetracycline is dependent mainly on *S. rimosus* and its mutant strains, but other species, including *S. platensis*, *S. vendargensis* and *S. gilvus*, have been reported to produce the antibiotic.

Streptomyces rimosus does not utilize sucrose, but suitable sources of carbohydrate include glucose, glycerol, lactose, maltose and starch. High-yielding mutant strains are able to metabolize some vegetable oils (e.g. soya bean oil and peanut oil). Oxytetracycline-producing strains are able to use nitrogen sources similar to those available to strains of *S. aureofaciens*.

8.2. *The chlortetracycline–tetracycline fermentation*

The ratio of the amounts of chlortetracycline and tetracycline produced is directly related to the concentration of chloride ions in the medium: the tetracycline content ranges from about 30% in the presence of 1·7 m-equiv./1. chloride ions to only about 7·5% when the content of chloride ions exceeds 6·4 m-equiv./1. Some commercial processes for production of tetracycline therefore depend on removal of chloride ions from the medium.

The proportion of tetracycline can be increased to a maximum of about 90% by including in the medium suitable inhibitors of the chlorination reaction: these include bromide ions (0·8–48 m-equiv./1.), thiocyanate and any of a considerable number of patented compounds, including many that are able to sequester copper, a metal that is essential for the chlorinating enzyme system:

- mercapto-dimethylthiazole (U.S.P. 2956930);
- thioanilide (U.S.P. 2949407);
- thiourea and certain of its derivatives (U.S.P. 2949406);
- *N*-cyclohexyl-2-benzothiazole sulphenamide (U.S.P. 2940909);
- certain sulphur-substituted diazine compounds (U.S.P. 2923668);
- thiocarbamates and related compounds (U.S.P. 2923667);
- 2-benzoxazolyl-bis-*N*-methylcarbamate or 2-chlorobenzoxazole or 2-chlorobenzothiazole (U.S.P. 2940906);
- 2,6-dimercapto-benzoxazole (U.S.P. 2940907);
- 2,5-disubstituted 1,3,4-thiadiazoles (B.P. 845715);
- certain pyridazine compounds (B.P. 836657);
- compounds containing the group $>N—C—S—$ (B.P. 846510).

Media for producing any of the tetracycline antibiotics are generally similar, but it must be remembered that strains of *S. rimosus* cannot use sucrose as carbohydrate source and that the ratio of chlortetracycline:tetracycline must be controlled by the appropriate methods already described. A typical medium, capable of yielding 2500 μg/ml chlortetracycline, would contain: cornsteep liquor, 3·0% w/v; starch, 1·5–8·0%; $CaCO_3$, 0·9%; $(NH_4)_2SO_4$, 0·33%; NH_4Cl, 0·1%; $MgCl_2,6H_2O$, 0·2%; $FeSO_4$, $4H_2O$, 0·005%; $ZnSO_4,8H_2O$, 0·01%; $CoCl_2,6H_2O$, 0·0005%; lard oil, 0·5–3·0% (added at intervals during the period of fermentation); pH to 6·6–6·8 before sterilization, to give an initial pH of 7·0.

The oil is considered to have a beneficial effect in limiting excessive alkalinity in the medium, besides its role as an anti-foam material. The magnesium and calcium salts included in the medium have a dual function: their ions aid regulation of the pH of the medium and also minimize the toxicity of the antibiotics to the production strains by saturating the sequestering power of the antibiotics for essential trace metals.

As in the streptomycin fermentation, the concentration of phosphate is critical for producing the tetracycline antibiotics: 0·03% w/v KH_2PO_4 diminishes chlortetracycline production by about 50%. It is therefore essential to ensure that phosphate, present in the cornsteep liquor, is not excessive. The mechanism of the phosphate effect in diminishing the yield of tetracycline antibiotics has not been determined, but it seems probable that regulation of the Warburg–Dickens oxidative pentose-phosphate and Embden–Meyerhof–Parnas glycolytic pathways for utilization of carbohydrate (see Chapter 9) may be involved.

Benzyl thiocyanate (0·0004% w/v) included in the medium has a particularly beneficial effect on the yield of chlortetracycline: its function is not known, but it has been suggested that its mode of action is to stimulate some unidentified stage in the biosynthesis of chlortetracycline particularly sensitive to even a transient deficiency of aeration.

Suggested literature for further reading

1. Arison, B. H., Wendler, N. L., Taub, D., Hoffsommer, R. D., Kuo, C. H., Slates, H. L. and Trenner, L. R., The delineation of griseofulvin and related systems by nuclear magnetic resonance spectroscopy, *J. Am. chem. Soc.* **85**, 627–31 (1963).
2. Ashton, G. C. and Brown, A. P., Determination of griseofulvin in fermentation samples. Part I. Spectrophotometric assay. (Appendix to Part I. Seven-point correction procedure. Ashton, G. C. and Tootill, J. P. R.), *Analyst Lond.* **81**, 220–7 (1956).

3. Ashton, G. C. and Tootill, J. P. R., A method of correcting for irrelevant absorption in ultra-violet spectrophotometric analysis, *Analyst Lond.* **81,** 232–8 (1956).
4. Barton, D. H. R. and Cohen, T., *Some Biogenetic Aspects of Phenol Oxidation*, Festschrift A. Stoll, Bickhauser, Basle, 1956.
5. Bentley, R., Biochemistry of fungi, *A. Rev. Biochem.* **31,** 589–624 (1962).
6. Birch, A. J. and Donovan, F. W., Studies in relation to biosynthesis. I. Some possible routes to derivatives of orcinol and phloroglucinol, *Aust. J. Chem.* **6,** 360–8 (1953).
7. Birch, A. J., Massy-Westropp, R. A. and Moye, C. J., Studies in relation to biosynthesis. VII. 2-Hydroxy-6-methylbenzoic acid in *Penicillium griseofulvum*, *Aust. J. Chem.* **8,** 539–44 (1955).
8. Birch, A. J., Cassera, A. and Rickards, R. W., Intermediates in biosynthesis from acetic acid units, *Chemy. Ind.* 792–3 (1961).
9. Birch, A. J., Some pathways in biosynthesis (Simonsen lecture), *Proc. chem. Soc. Lond.* 3–13 (1962).
10. Birch, A. J., Massy-Westropp, R. A., Rickards, R. W. and Smith, H., Studies in relation to biosynthesis. Part XIII. Griseofulvin, *J. chem. Soc.* 360–5 (1958).
11. Birch, A. J., English, R. J., Massy-Westropp, R. A. and Smith, H., Studies in relation to biosynthesis. Part XV. Origin of terpenoid structures in mycelianamide and mycophenolic acid, *J. chem. Soc.* 369–75 (1958).
12. Birch, A. J., Snell, J. F. and Thompson, P. J., Studies in relation to biosynthesis. Part XXVIII. Oxytetracycline (Terrramycin), *J. chem. Soc.* 425–9 (1962).
13. Brossi, von A., Baumann, M. and Burkhardt, F., Syntheseversuche in der Griseofulvinreihe. 5. Zur Synthese von (—)-und (+)-Griseofulvin und ihrer Diastereomeren, *Helv. Chim. Acta* **45,** 1292–7 (1962).
14. Bu'Lock, J. D. and Smalley, H. M., Biosynthesis of aromatic substances from acetyl- and malonyl-coenzyme A, *Proc. chem. Soc. Lond.* 209–11 (1961).
15. Bu'Lock, J. D., Smalley, H. M. and Smith, G. N., Malonate as a biosynthetic intermediate in *Penicillium urticae*, *J. biol. Chem.* **237,** 1778–80 (1962).
16. Bu'Lock, J. D., Intermediary metabolism and antibiotic synthesis, *Adv. appl. Microbiol.* **3,** 293–342 (1961).
17. Cathou, R. E. and Buchanan, J. M., Enzymatic synthesis of the methyl group of methionine. V. Studies with 5,10-methylenetetrahydrofolate reductase from *Escherichia coli*, *J. biol. Chem.* **238,** 1746–51 (1963).
18. Daly, C., A simplified spectrophotometric assay of griseofulvin in fermentation samples, *Analyst Lond.* **86,** 129–31 (1961).
19. Day, A. C., Nabney, J. and Scott, A. I., Oxidative pairing of phenolic radicals. Part I. The total synthesis of griseofulvin, *J. chem. Soc.* 4067–74 (1961).

20. DEAN, F. M., Biosynthesis (Chapter 18 in *Naturally Occurring Oxygen Ring Compounds*, 586–618), Butterworths, London, 1963.
21. DOREY, M. J., MITCHELL, I. L. S., RULE, D. W. and WALKER, C., *Production of Griseofulvin*, British Patent, 934528 (1960).
22. DULANEY, E. L., PUTTER, I., DRESCHER, D., CHAIET, L., MILLER, W. J., WOLF, F. J. and HENDLIN, D., Transethylation in antibiotic biosynthesis. I. An ethyl homolog of oxytetracycline, *Biochim. biophys. Acta* **60,** 447–9 (1962).
23. FOSTER, M. A., TEJERINA, G., GUEST, J. R. and WOODS, D. D., Two enzymic mechanisms for the methylation of homocysteine by extracts of *Escherichia coli*, *Biochem. J.* **92,** 476–88 (1964).
24. FRIEDKIN, M., Enzymatic aspects of folic acid, *A. Rev. Biochem.* 185–214 (1963).
25. GATENBECK, S., The biosynthesis of oxytetracycline, *Biochem. biophys. Res. Commun.* **6,** 422–6 (1962).
26. GOODMAN, J. J. and MILLER, P. A., The effect of antimetabolites on the biosynthesis of tetracyclines, *Biotechnology and Bioengineering* **4,** 391–402 (1962).
27. GOODMAN, J. J., *Streptomyces Aureofaciens Fermentation Process*, U.S.A. Patent, 2911339 (1959).
28. GOSHTYALEK, Z., YANECHEK, YU, DOSKOCIL, YU and KASHPAROVA, I., The influence of interruption in aeration, and of orthophosphates and benzyl thiocyanate on the formation of chlortetracycline, *Antibiotiki* **4,** 34–6 (1959).
29. GREENBERG, D. M., Biological methylation, *Adv. Enzymol.* **25,** 395–431 (1963).
30. GUEST, J. R., FOSTER, M. A. and WOODS, D. D., Transfer of the methyl group from N^5-methyl-tetrahydrofolate to homocysteine in *Escherichia coli*, *Biochem. J.* **92,** 497–504 (1964).
31. GUEST, J. R., FOSTER, M. A. and WOODS, D. D., Methyl derivatives of folic acid as intermediates in the methylation of homocysteine by *Escherichia coli*, *Biochem. J.* **92,** 488–96 (1964).
32. HOCHSTEIN, F. A., VON WITTENAU, M. S., TANNER, F. W. Jr. and MURAI, K., 2-Acetyl-decarboxamidooxytetracycline, *J. Am. chem. Soc.* **82,** 5934–7 (1960).
33. HOCKENHULL, D. J. D. and FAULDS, W. F., Origin of the methoxy groups of griseofulvin, *Chemy. Ind.* 1390 (1955).
34. HOCKENHULL D. J. D., *Improvements in or Relating to Antibiotics*, British Patent, 868958 (1959).
35. KISLUIK, R. L., The source of hydrogen for methionine methyl formation. *J. biol. Chem.* **238,** 397–400 (1963).
36. LARRABEE, A. R., ROSENTHAL, S., CATHOU, R. E. and BUCHANAN, J. M., Enzymatic synthesis of the methyl group of methionine. IV. Isolation, characterization, and role of 5-methyltetrahydrofolate, *J. biol. Chem.* **238,** 1025–31 (1963).
37. LEIN, J., SAWMILLER, L. F. and CHENEY, L. C., Chlorination inhibitors affecting the biosynthesis of tetracycline, *Appl. Microbiol.* **7,** 149–51 (1959).

38. LOUGHLIN, R. E., ELFORD, H. L. and BUCHANAN, J. M., Enzymatic synthesis of the methyl group of methionine. VII. Isolation of a cobalamin-containing transmethylase (5-methyltetrahydrofolate-homocysteine) from mammalian liver, *J. biol. Chem.* **239** (9) 2888–95 (1964).
39. MCCORMICK, J. R. D., SJOLANDER, N. O., HIRSCH, U., JENSEN, E. R. and DOERSCHUK, A. P., A new family of antibiotics: the demethyl-tetracyclines, *J. Am. chem. Soc.* **79**, 4561–3 (1957).
40. MCCORMICK, J. R. D., JOHNSON, S. and SJOLANDER, N. O., Biosynthesis of the tetracyclines. V. Naphthacenic precursors, *J. Am. chem. Soc.* **85**, 1692–4 (1963).
41. MCCORMICK, J. R. D., MILLER, P. A., JOHNSON, S., ARNOLD, N. and SJOLANDER, N. O., Biosynthesis of the tetracyclines. IV. Biological re-hydration of the 5a 6-anyhdrotetracyclines, *J. Am. chem. Soc.* **84**, 3023–5 (1962).
42. MCCORMICK, J. R. D., MILLER, P. A., GROWICH, J. A., SJOLANDER, N. O. and DOERSCHUK, A. P., Two new tetracycline-related compounds: 7-chloro-5a(11a)-dehydrotetracycline and 5a-epi-tetracycline. A new route to tetracycline, *J. Am. chem. Soc.* **80**, 5572–3 (1958).
43. MCCORMICK, J. R. D., SJOLANDER, N. O., MILLER, P. A., HIRSCH, U., ARNOLD, N. and DOERSCHUK, A. P., The biological reduction of 7-chloro-5a (11a)-dehydrotetracycline to 7-chloro-tetracycline by *Streptomyces aureofaciens*, *J. Am. chem. Soc.* **80**, 6460–1 (1958).
44. MCCORMICK, J. R. D., JOACHIM, U., HIRSCH, U., JENSEN, E. R., JOHNSON, S. and SJOLANDER, N. O., Biosynthesis of the tetracyclines. VII. 4-hydroxy-6-methlypretetramid, an intermediate accumulated by a blocked mutant of *Streptomyces aureofaciens, J. Am. chem. Soc.* **87**, 1793–4 (1965).
45. MCCORMICK, J. R. D., *Cosynthetic Factor*-1, *and its Production*, U.S.A. Patent, 2996499 (1961).
46. MCCORMICK, J. R. D. and GARDNER, W. E., *Naphthacenequinones*, U.S.A. Patent, 3074975 (1963).
47. MCCORMICK, J. R. D., SJOLANDER, N. O. and HIRSCH, U., *Biological Conversion of* 5*a* (11*a*)*-dehydrotetracyclines to Broad-spectrum Antibiotics*, U.S.A. Patent, 2965546 (1960).
48. MCGHEE, W. J. and MEGNA, J. A., *Process for Production of Tetracycline*, U.S.A. Patent, 2776243 (1957).
49. MCMASTER, W. J., SCOTT, A. I. and TRIPPETT, S., Metabolic products of *Penicillium patulum*, *J. chem. Soc.* 4628–31 (1960).
50. MACMILLEN, J., Griseofulvin. VII. Dechlorogriseofulvin, *J. chem. Soc.* 1697–1702 (1953).
51. DI MARCO, A. and PENNELLA, P., The fermentation of the tetracyclines, *Prog. ind. Microbiol.* **1**, 47–92 (1959).
52. MILLER, P. A., SJOLANDER, N. O., NALESNYK, S., ARNOLD, N., JOHNSON, S., DOERSCHUK, A. P. and MCCORMICK, J. R. D., Cosynthetic Factor I. a factor involved in hydrogen-transfer in *Streptomyces aureofaciens, J. Am. chem. Soc.* **82**, 5002–3 (1960).
53. MINIERI, P. P., FIRMAN, M. C. and SOKOL, H., *Deionized Corn Steep Liquor in Production of Tetracycline*, U.S.A. Patent, 2866738 (1958).

54. NIEDERCORN, J. G., *Process for Producing Aureomycin*, U.S.A. Patent, 2609329 (1952).
55. PETTY, M. A. Jr., *Production of Chlortetracycline*, U.S.A. Patent, 2709672 (1955).
56. RHODES, A., Griseofulvin: production and biosynthesis, *Prog. ind. Microbiol.* **IV**, 166–87 (1962).
57. RHODES, A. and MCGONAGLE, M. P., *Antibiotic Production*, British Patent, 934526 (1963).
58. RHODES, A., BOOTHROYD, B., MCGONAGLE, M. P. and SOMERFIELD, G. A., Biosynthesis of griseofulvin: the methylated benzophenone intermediates, *Biochem. J.* **81**, 28–37 (1961).
59. RHODES, A., SOMERFIELD, G. A. and MCGONAGLE, M. P., Biosynthesis of griseofulvin. Observations on the incorporation of (^{14}C) griseophenone C and (^{36}Cl) griseophenones B and A, *Biochem. J.* **88**, 349–57 (1963).
60. RHODES, A., CROSSE, R., FERGUSON, T. P. and FLETCHER, D. L., *Improvements in or Relating to the Production of the Antibiotic Griseofulvin*, British Patent, 784618 (1955).
61. ROWBURY, R. J. and WOODS, D. D., O-succinylhomoserine as an intermediate in the synthesis of cystathionine by *Escherichia coli*, *J. gen. Microbiol.* **36**, 341–58 (1964).
62. SCOTT, A. I. and BEDFORD, C. T., Simulation of the biosynthesis of tetracyclines. A partial synthesis of tetracycline from anhydroaureomycin, *J. Am. chem. Soc.* **84**, 2271–2 (1962).
63. SHAPIRO, S. K., and SCHLENK, F., The biochemistry of sulfonium compounds, *Adv. in Enzymol.* **22**, 237–80 (1960).
64. SHAW, P. D. and HAGER, L. P., Biological chlorination. IV. Chloroperoxidase: a component of the β-ketoadipate chlorinase system, *J. biol. Chem.* **236**, 1626–30 (1961).
65. SNELL, J. F., BIRCH, A. J. and THOMPSON, P. L., The biosynthesis of tetracycline antibiotics, *J. Am. chem. Soc.* **82**, 2402 (1960).
66. SOBIN, B. A., FÎNLAY, A. C. and KANE, J. H., *Terramycin and its Production*, U.S.A. Patent, 2516080 (1950).
67. THOMAS, R., Studies in the biosynthesis of fungal metabolites. 2. The biosynthesis of alternariol and its relation to other fungal phenols, *Biochem. J.* **78**, 748–58 (1961).
68. WEISSBACH, H., PETERKOFSKY, A., REDFIELD, B. G. and DICKERMAN, H., Studies on the terminal reaction in the biosynthesis of methionine, *J. biol. Chem.* **238**, 3318–24 (1963).
69. WHALLEY, W. B., *Recent Developments in the Chemistry of Natural Phenolic Compounds*, Pergamon, London, 1961.
70. WRIGHT, L. D., Biosynthesis of isoprenoid compounds, *A. Rev. Biochem.* 525–48 (1962).

CHAPTER 15

Vitamin B_{12}: Structure, Biosynthesis and Production

VITAMIN B_{12} is one of the water-soluble vitamins necessary, in minute amount (0·001 mg per day), to prevent pernicious anaemia in man. Vitamin B_{12} is an essential growth factor for some bacteria, but it is synthesized in small quantity by many other micro-organisms, and some of these are used in industrial microbiological processes yielding up to about 20μg/ml.

I. Structure of vitamin B_{12}

The molecule of vitamin B_{12} ($C_{63}H_{88}O_{14}N_{14}PCo$) has a complex structure (Fig. 15.1), comprising a planar group linked through 1-amino-2-propanol with a "nucleotide" in a plane almost at right-angles. The planar group has a central atom of cobalt in co-valent linkages with a cyanide group and with one of the four reduced pyrrole rings of a linear tetrapyrrole derivative that has been folded into a macro-ring: additional co-ordinate bonds link the cobalt atom to the other three reduced pyrrole rings. Methine groups (= CH —) form three of the bridges between the pyrrole rings, as in the porphyrins, but in vitamin B_{12} the macro-ring is completed by direct linkage of the α-carbon of ring *A* with the corresponding carbon of ring *D*. Eleven of the nineteen carbon atoms forming the macro-ring carry substituent methyl, acetamido or propionamido groups.

The nucleotide of the vitamin B_{12} molecule also is not typical: the base is 5,6-dimethylbenziminazole instead of the purine or

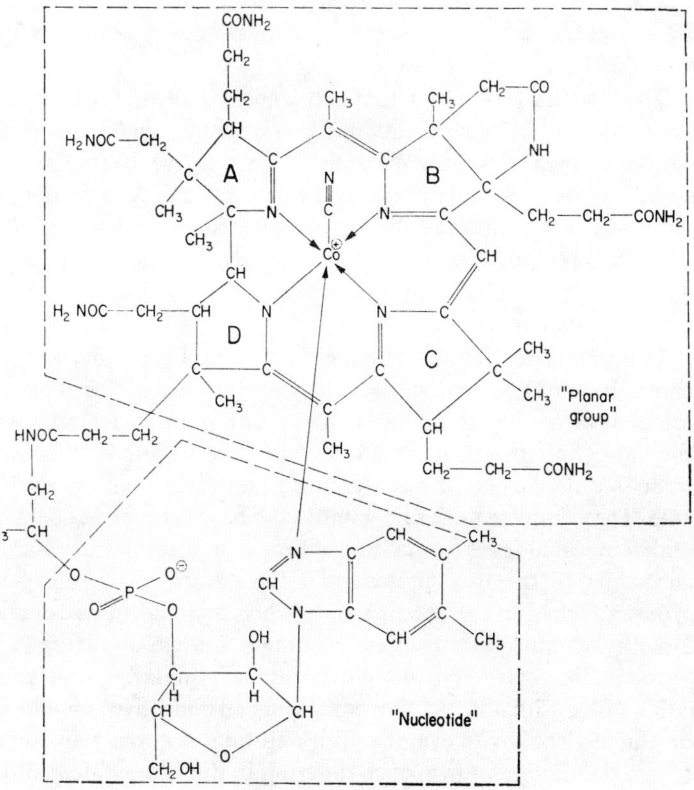

FIG. 15.1. Structure of vitamin B_{12}.

pyrimidine of nucleic acids, and the glycosidic link with ribose is α- instead of the usual β-linkage in nucleic acids. The phosphate group of the nucleotide is esterified with 1-amino-2-propanol, and this provides one of the bridges to the planar group. The other bridge between the two portions of the molecule is formed by a co-ordinate bond between the central cobalt atom of the planar group and one of the nitrogen atoms of the benziminazole base in the nucleotide.

The cobalt atom carries a positive charge, but this is neutralized by the negative charge on the phosphoric acid residue, so that vitamin B_{12} is an inner salt.

The cyanide group of vitamin B_{12} may be readily removed by photolysis or by suitable reducing conditions, and the residual cation is then termed cobalamin. The positive charge on the cobalt atom of cobalamin may be neutralized by a variety of anions, and the nature of the ion is indicated by the nomenclature adopted for these analogues: for example, vitamin B_{12} (cyanide anion) is cyanocobalamin and vitamin B_{12a} (hydroxyl anion) is hydroxocobalamin.

The acidic nucleotide residue may also be removed under carefully controlled conditions, leaving the rest of the molecule intact as factor B (cobinamide). This reaction can be reversed with the aid of micro-organisms and can be used to link with cobinamide various nucleotides containing bases other than benziminazole. The complete molecule without the base is termed cobamide, and these analogues containing different nucleotides are named according to the base present and the nature of the inorganic group attached to the cobalt atom. When the nucleotide contains 2-methyladenine, the resultant analogue, known as factor A, is termed 2-methyladenylcobamide cyanide: similarly, linkage of cobinamide with a nucleotide containing adenine gives pseudo-B_{12} or α-adenylcobamide cyanide. Ten naturally occurring analogues of cobalamin are known, each differing in the base content of the nucleotide, but cobalamin alone is clinically effective. Many of these analogues were first discovered as products of micro-organisms active in sewage.

2. Biosynthesis of cobalamin

The main building-block for the macro-ring of the planar group is δ-aminolevulinic acid formed by condensation of succinyl-coenzyme A with glycine to give succinylglycine, which then loses the carboxyl group of the glycine moiety as carbon dioxide:

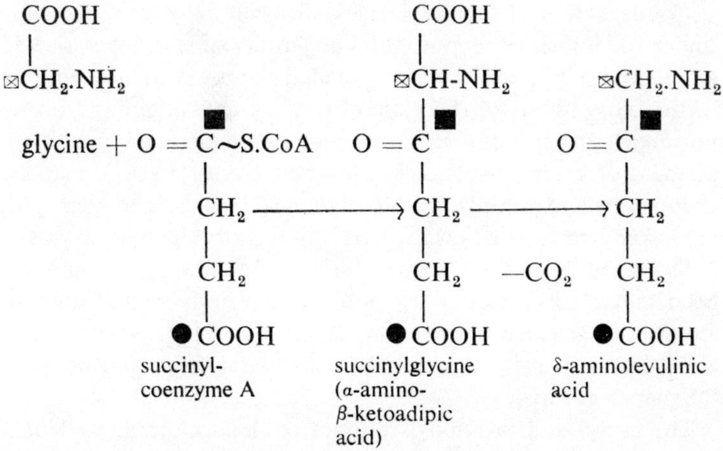

This biosynthesis is catalysed by pyridoxal phosphate and may also require biotin as a co-factor.

Condensation of two molecules of δ-aminolevulinic acid gives porphobilinogen:

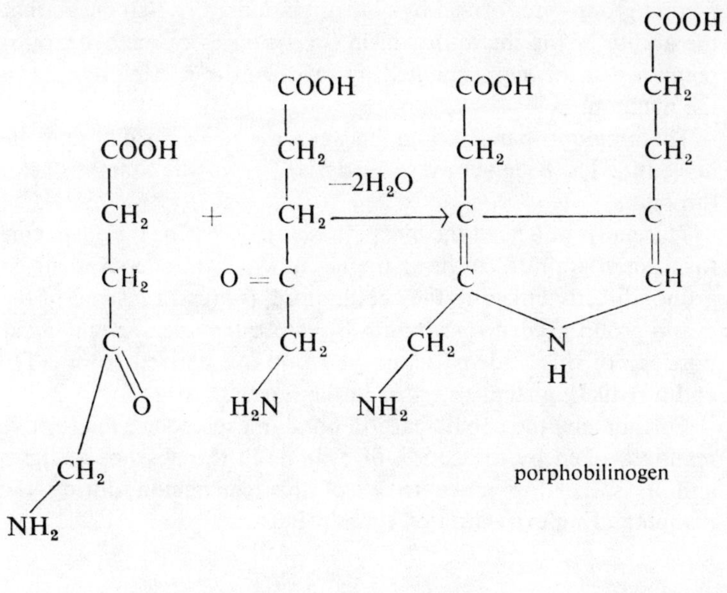

Condensation of four molecules of porphobilinogen, probably under the influence of porphobilinogen deaminase, gives rise to a linear tetrapyrrole. This intermediate appears to be common to the biosynthesis of the planar group of cobalamin and of the porphyrin system found in chlorophyll (the magnesium porphyrin pigment of green leaves), in heme (the iron porphyrin prosthetic group of haemoglobin) and in the cytochromes (the iron porphyrin oxido-reductases associated with the respiratory chain of enzymes).

Details of the subsequent biosynthesis are not known, but it has been suggested that the linear tetrapyrrole folds so that the two terminal rings close to form a macro-ring and that isomerization aided by an isomerase then occurs to give a stable precursor of the planar group.

The acetyl and propionyl residues of the acetamido and propionamido side-chains, respectively, are provided by the porphobilinogen molecules that form the macro-ring, but it is probable that ATP, coenzyme A or possibly both substances will be necessary in addition to ammonia for biosynthesis of the numerous amide bonds. It would also seem probable that most of the methyl groups are formed by *C*-methylations: this may mean that the ability of the micro-organism for synthesis of methyl groups may need to be supplemented by provision of methyl donors in the medium.

The aminopropanol bridge between the planar group and the nucleotide has been shown to be derived from the α-amino acid, threonine.

The end-products of the biosynthesis appear to be the coenzyme forms of vitamin B_{12} or its analogues in which a 5′-deoxyadenosyl group, directly linked to the cobalt atom, is present instead of the cyano group needed for vitamin B_{12}. Both the adenine and sugar portions of this 5′-deoxyadenosyl group are derived from ATP, and a reduction step is needed in the biosynthesis.

Fortunately, the cobalt–carbon bond in each coenzyme form is readily broken by the action of cyanide in the absence of light, and it is usual practice to effect this conversion during the manufacturing extraction of vitamin B_{12}.

3. Production of vitamin B_{12}

Vitamin B_{12} can be obtained as a by-product during production of streptomycin by *Streptomyces griseus* strains or of chlortetracycline by strains of *S. aureofaciens*, but yields do not usually exceed 2μg/ml. Pharmacopoeial grade vitamin B_{12} is usually obtained more economically, and in yield of up to about 20 μg/ml, by bacterial processes involving a strain of *Propionibacterium freudenreichii* or of *Pseudomonas* species to produce the vitamin as the major useful product of fermentation.

Some of the micro-organisms used for commercial production of vitamin B_{12} are unable to synthesize efficiently the benziminazole ring system of the nucleotide. *Pb. freudenreichii* forms cobinamide (factor B); if 5,6-dimethylbenziminazole or the corresponding phenylenediamine is supplied in the later stages of the process as a pre-formed intermediate, the micro-organism readily incorporates the base into cobalamin. It is also usual to include a methyl donor (e.g. betaine) in the medium, to overcome any deficiency in the capacity of the micro-organism for biosynthesis of methyl groups.

A suitable medium for production of vitamin B_{12} by *Pb. freudenreichii* would contain: glucose 10% w/v; cornsteep liquor to give 0·32% w/v nitrogen; betaine 0·5–1% w/v (supplied as a soluble salt); cobalt 3–5 μg/ml (e.g. as $CoCl_2,6H_2O$); adjusted to give a post-sterilization pH of about 7·0–7·5.

It is preferable to conduct the early part of the fermentation process for production of cobinamide under substantially anaerobic conditions, but completion of the cobalamin biosynthesis requires a low rate of aeration from about 70 hr until harvest at about 120 hr, with simultaneous addition of 5,6-dimethylbenziminazole (about 0·01%) to the fermentation broth.

The pH of the fermentation broth falls to 6·5–7·0 during the first 20 hr, owing to rapid growth of the culture, with concomitant production of organic acids. Thereafter the pH rises slowly to about 8·5 at the end of the fermentation period. Some processes may require addition of ammonia in order to maintain a slightly alkaline pH.

The cobalamins bound within the cells are released by acidifying the harvest broth to pH 2–3; sometimes it may also be necessary to boil the broth for 5 min. Finally cobalamins (total B_{12}) are all converted to cyanocobalamin (true B_{12}) by addition of potassium cyanide during the extraction process.

Suggested literature for further reading

1. BLAYLOCK, B. A. and STADTMAN, T. C., Enzymic formation of methylcobalamin in *Methanosarcina Barkerii* extracts, *Biochem. biophys. Res. Commun.* **17** (5) 475–80 (1964).
2. BONNETT, R., Chemistry of the vitamin B_{12} group, *Chem. Rev.* **63,** 573–605 (1963).
3. CORNFORD, P., Transformation of porphobilinogen into porphyrins by preparations from human erythrocytes, *Biochem. J.* **91,** 64–73 (1964).
4. DOLPHIN, D., JOHNSON, A. W., RODRIGO, R. and SHAW, N., The vitamin B_{12} coenzyme, *Proceedings of the Symposium on the Chemistry and Biochemistry of Fungi and Yeasts*, Dublin, 1963, Butterworths, London, 1963.
5. JUKES, T. H. and BROQUIST, H. P., Biogenesis and metabolism of folic acid and vitamin B_{12} (Chapter 24 in *Metabolic Pathways II*), 713–40, Academic Press, New York and London, 1961.
6. KNIVETT, V. A., The Microbiological production of vitamin B_{12} and sulphide from sewage, *Prog. ind. Microbiol.* **2,** 27–45 (1960).
7. LASCELLES J., Tetrapyrrole synthesis in micro-organisms (Chapter 7 in *The Bacteria. III: Biosynthesis*), 335–72, Academic Press, New York and London, 1962.
8. MERVYN, L. and SMITH, E. L., The biochemistry of Vitamin B_{12} fermentation, *Prog. ind. Microbiol.* **5,** 151–201 (1964).
9. SMITH, E. L., *Vitamin B_{12}* (3rd edition), Methuen's Monographs on biochemical subjects, Methuen, London, 1965.
10. SPEEDIE, J. D. and HULL, G. W., *Cobalamin Producing Fermentation Process*, British Patent 829232.

CHAPTER 16

Oxidative Disposal of Sewage, Industrial Wastes and Aromatic Compounds by Micro-organisms

MODERN methods for disposal of sewage and industrial wastes frequently depend upon fermentation processes. The fundamental requirements are to remove objectionable solids, and to lower the biochemical oxygen demand (B.O.D.) defined as the milligrams of oxygen that would be utilized in a specified period of time (usually 5 days) during microbiological oxidation at a specified temperature (usually 20°C) of the organic matter in one litre of the waste.

Direct disposal of wastes having high B.O.D. values to surface waters and streams is precluded because of the danger to aquatic life and public health. Water-borne disease organisms present in waters contaminated with sewage may result in epidemic outbreaks, but depletion of the dissolved oxygen content of streams may be almost as serious. The dissolved oxygen essential for aquatic life in surface waters is limited in amount, but it is maintained in an equilibrium state by normal processes of physical aeration. Influx of readily fermentable sewage and industrial wastes could cause rapid proliferation of micro-organisms with concomitant consumption in respiration of the available oxygen at a rate greater than that at which it can be replenished by natural aeration. The resultant anaerobic conditions would result in loss of aquatic life and generation of the disagreeable odours that often accompany the anaerobic microbiological decomposition of organic matter.

We have seen in Chapter 8 that many industrial wastes can be eliminated by utilizing them as substrates for production of food- and fodder yeasts or for production of industrial alcohol by fermentation: these include acid hydrolysates of wood and other cellulosic materials from wood pulp and paper industries, whey from dairy wastes, surplus grains from agriculture, and the waste wax content of kerosenes and gas oils from the petroleum industry. Modern methods of sewage and industrial waste disposal frequently depend upon aerobic fermentations by mixed populations of bacteria, protozoa and, to a lesser extent, fungi, but the principle of operation is essentially similar in all, although a variety of mechanical techniques may be employed. Certain industries provide wastes containing characteristic organic materials that may sometimes present additional problems: these include the detergents from laundries and the organic substances from factories producing industrial chemicals.

The general principles of modern waste disposal are briefly considered below, but it should be realized that our knowledge of their biochemistry is incomplete. Many of the detailed mechanisms for microbiological degradation of substrate carbohydrates, fats and nitrogenous materials in respiration and growth are clear, but relatively little is known of the routes involved in oxidation of the various aromatic substances that can occur in industrial effluents. Rapid progress is now being made in this field, and it seems probable that the information obtained may eventually provide less empirical microbiological methods for the more difficult aspects of waste disposal.

1. Oxidation of aromatic compounds by micro-organisms

Degradation of aromatic compounds appears to depend on formation of inducible enzymes (see Chapter 8): most of our knowledge is based upon studies with bacteria, particularly soil pseudomonads, but it is probable that this will be extended by investigations into the degradation by fungi of the "humic acid" of soil.

1.1. *Preliminary transformations to key intermediates*

Present knowledge indicates that various aromatic compounds undergo preliminary transformations to several key intermediates, including catechol (I), protocatechuic acid (II), kynurenic acid (III) and probably homogentisic (IV) and gentisic (V) acids. The nature of the key intermediate is largely determined by the structure of the substrate to be degraded:

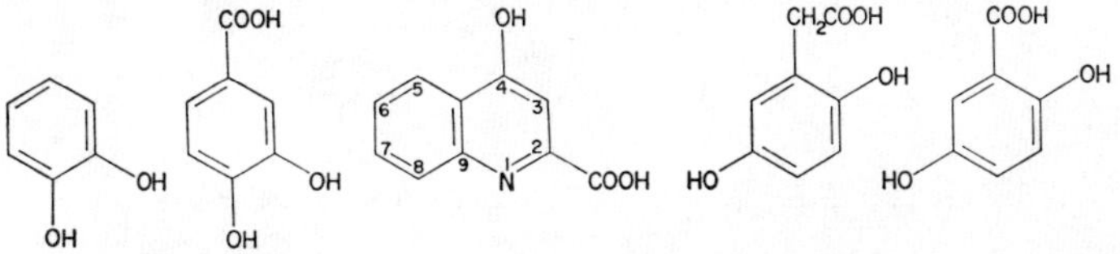

(I) (II) (III) (IV) (V)

The preliminary transformations occurring in the degradation of naphthalene by soil pseudomonads (Fig. 16.1) is a process that has been studied in detail and will serve to exemplify some of the general principles involved. Oxidation of naphthalene (VI) gives first D-trans-1,2-dihydro-1,2-dihydroxynaphthalene (VII), and then 1,2-dihydroxynaphthalene (VIII), the substrate for the subsequent ring cleavage; *cis-ortho*-hydroxybenzalpyruvate (IX and X) is the immediate product of this ring fission at the bond adjacent to the hydroxyl groups between the angular carbon atom and C-1 of the naphthalene nucleus. The enzyme involved is an oxygenase dependent on ferrous ions. Some of the *trans* form is also made by isomerization. Both isomers are converted by an aldolase type enzyme to salicylaldehyde (XI) and a C_3 residue (probably pyruvic acid). The salicylaldehyde is oxidized by a NAD-specific dehydrogenase to salicylic acid (XII), and the preliminary transformations are completed when this intermediate is oxidatively decarboxylated to catechol.

Preliminary transformations occurring in various other benzenoid compounds, but terminating in production of protocatechuic acid, are shown in Fig. 16.2. Oxidation of the side-chain of these benzenoid compounds seems to depend on the presence of a *para*-hydroxyl group, but the process is unaffected by further ring

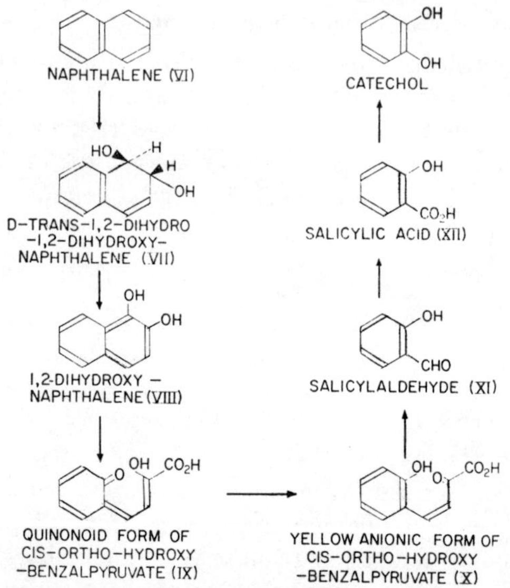

FIG. 16.1. Preliminary microbiological transformation of naphthalene to catechol in preparation for complete oxidation.

substitution of hydroxyl or methyl groups. Complete oxidation is, however, hindered by these additional groups, because the substituents inhibit the final ring hydroxylation necessary to complete the preliminary transformation to the key intermediate, protocatechuic acid.

Further metabolic facilities for preliminary degradation are exemplified in Fig. 16.3 by the transformation of the indole derivative, trytophan (XIII): cleavage of the five-membered heterocyclic ring provides a side-chain that is either oxidized by the aromatic pathway, yielding anthranilic acid and then catechol, or oxidatively deaminated by the quinoline pathway before recyclization to kynurenic acid.

Detergents, primarily composed of alkylbenzene sulphonates (XIV, R = alkyl) pose a special problem in effluent disposal: their resistance to microbial attack progressively increases with increased branching in the alkyl side-chain:

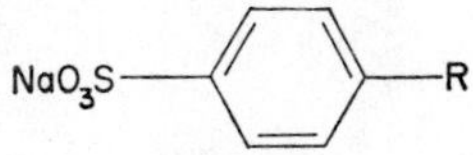

(XIV)

When the preliminary transformations to key intermediates are complete a relatively small number of oxidative pathways will suffice to continue the degradations, with formation of compounds that can be completely oxidized to carbon dioxide and water by the citrate cycle. Considerable progress has been made recently in elucidation of these oxidative pathways, and some of these will now be described.

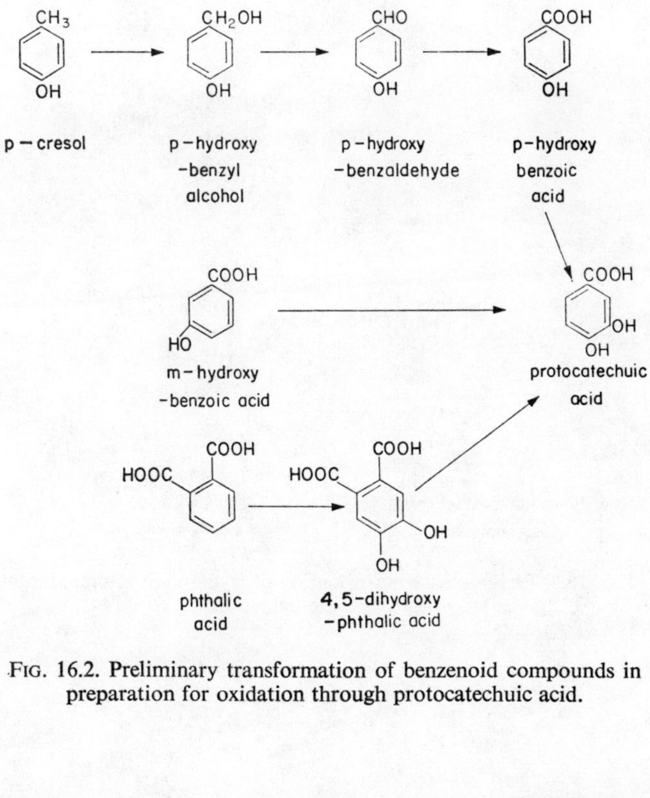

FIG. 16.2. Preliminary transformation of benzenoid compounds in preparation for oxidation through protocatechuic acid.

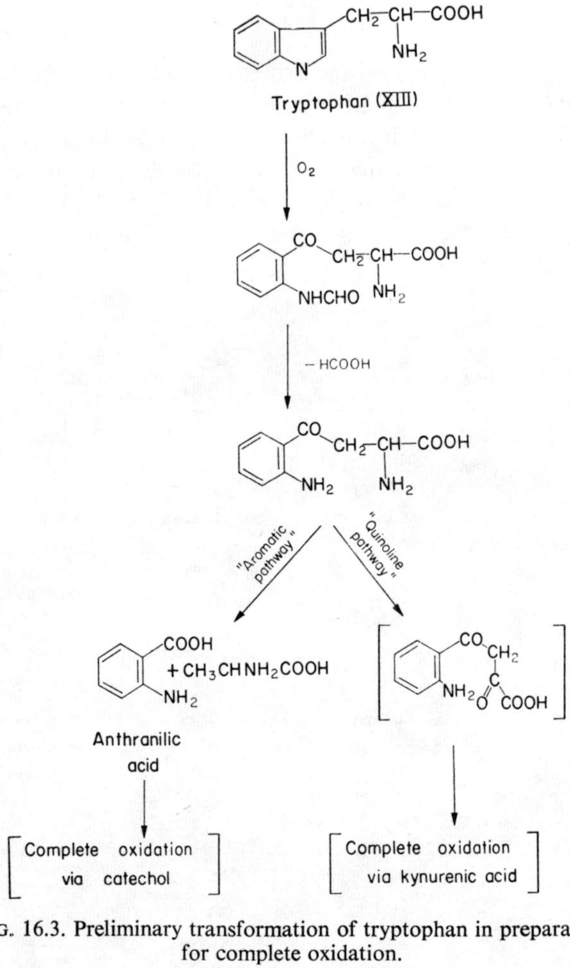

FIG. 16.3. Preliminary transformation of tryptophan in preparation for complete oxidation.

1.2. *Oxidation of catechol (I) and protocatechuic acid (II)*

Protocatechuic acid and catechol undergo oxidative ring fission by two alternative mechanisms (Fig. 16.4). Protocatechuate

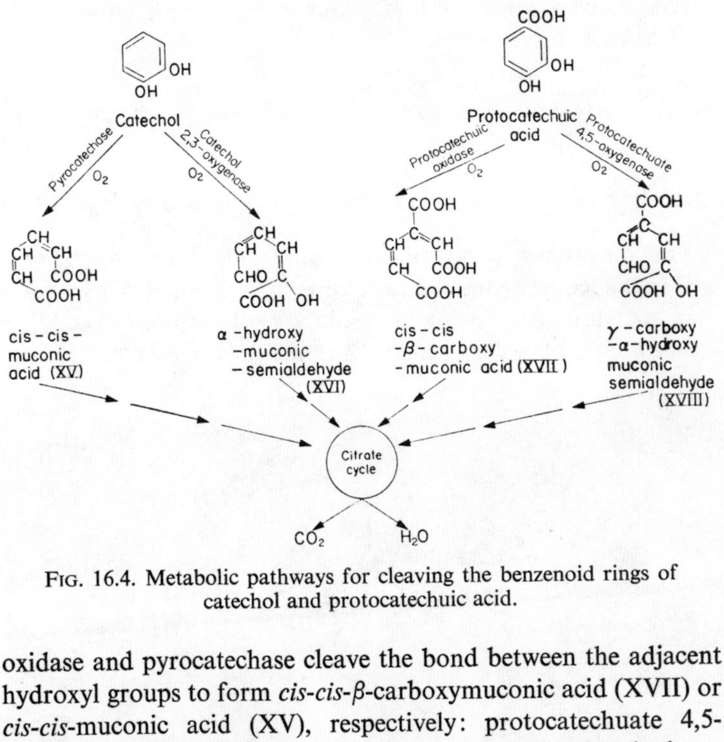

FIG. 16.4. Metabolic pathways for cleaving the benzenoid rings of catechol and protocatechuic acid.

oxidase and pyrocatechase cleave the bond between the adjacent hydroxyl groups to form *cis-cis-β*-carboxymuconic acid (XVII) or *cis-cis*-muconic acid (XV), respectively: protocatechuate 4,5-oxygenase and catechol 2,3-oxygenase, on the other hand, cleave the ring at the bond adjacent to that between the two hydroxyl groups affording γ-carboxy-α-hydroxymuconic semialdehyde (XVIII) and α-hydroxymuconic semialdehyde (XVI), respectively.

The aliphatic compounds resulting from cleavage of the benzenoid rings are then further metabolized by alternative routes to fit them for complete oxidation to carbon dioxide and water by the citrate cycle. Muconic and β-carboxymuconic acids both yield β-ketoadipic acid; in the presence of catalytic amounts of succinyl-coenzyme A, further conversion occurs to acetyl-coenzyme A and succinate, substrates for the citrate cycle:

(i) succinyl ~ S.Co$\bar{A}$ + β-ketoadipate $\xrightarrow{\text{thiophorase}}$ β-ketoadipyl ~ S.Co$\bar{A}$ + succinate

(ii) β-ketoadipyl ~ S.Co$\bar{A}$ + HS.Co$\bar{A}$ $\xrightarrow{\text{thiolase}}$ acetyl ~ S.Co$\bar{A}$ + succinyl S.Co$\bar{A}$

Overall: β-ketoadipate + HS.Co$\bar{A}$=acetyl ~ S.Co$\bar{A}$ + succinate

On the other hand, α-hydroxymuconic and γ-carboxy-α-hydroxymuconic semialdehydes appear to be attacked by most soil pseudomonads without evolution of carbon dioxide or consumption of oxygen according to the general scheme:

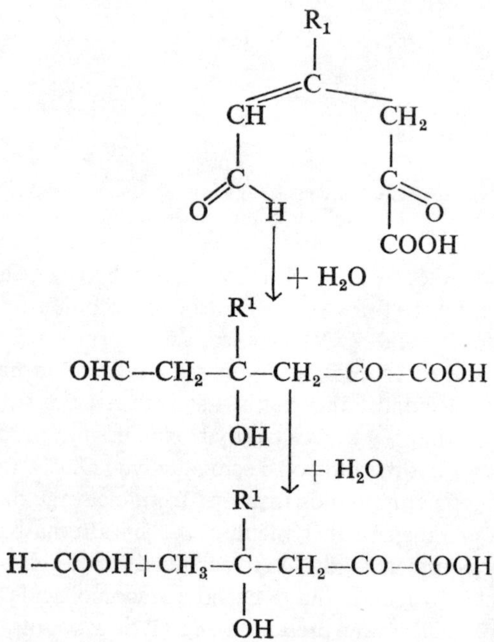

Formate is a common product from both substrates but the second product depends on the nature of R^1: α-hydroxymuconic semialdehyde ($R_1 = H$) gives α-keto-γ-hydroxyvaleric acid but γ-carboxy-α-hydroxymuconic semialdehyde ($R_1 = COOH$) yields γ-methyl-γ-hydroxy-α-ketoglutamic acid. Further degradation in the presence of $Mg^{\oplus\oplus}$ ions then yields compounds that can be completely oxidized by the citrate cycle: α-keto-γ-hydroxyvaleric acid forms an equimolar mixture of acetaldehyde and pyruvic acid, but γ-methyl-γ-hydroxy-α-ketoglutaric acid yields 2 moles of pyruvic acid.

1.3. *Oxidation of homogentisic acid (IV)*

The oxidation of homogentisic acid, a key intermediate in the bacterial degradation of tyrosine, phenylalanine and phenylacetic acid, is, at least in some micro-organisms, a single-step addition of oxygen. The reaction is catalysed by homogentisate oxidase to yield 4-maleylacetoacetate. Isomerization gives 4-fumarylacetoacetate, and this, in turn, is split to form fumarate and acetoacetate, substrates that can be completely oxidized to carbon dioxide and water via the citrate cycle.

1.4. *Oxidation of kynurenic acid (III)*

Completion of the first stage in this process yields the diol, 7,8-dihydro-7,8-dihydroxykynurenic acid, which is then dehydrogenated to 7,8-dihydroxykynurenic acid. Cleavage of the benzene ring occurs at the bond α to the two adjacent hydroxyl groups in a manner analogous to that already described for splitting the ring in 1,2-dihydroxynaphthalene (see Fig. 16.1), and then the pyridine ring is opened. The ultimate products are L-glutamic acid, D- and L-alanine, acetic acid and carbon dioxide but details of the pathways concerned have not yet been described.

2. Procedures for waste disposal

Most wastes, including domestic sewage, are transported by water in a closed system of pipes or sewers to the waste treatment plant. It is usual to ensure that the 5 day B.O.D. of the organic matter in the waste entering the municipal sewage treatment plant does not exceed about 200 mg/l. This normally presents no problem with ordinary municipal wastes, but industrial wastes, which frequently lie within the 5 day B.O.D. range of 1000–10,000 mg/l., must be diluted, or the rate of supply to the treatment plant must be regulated. Alternatively, an activated sludge plant that can operate at higher B.O.D. levels may be employed.

Purification of the carriage water to a degree permitting its disposal to surface waters depends upon the free play of the metabolic activities of micro-organisms, and this results in their rapid growth. Optimum conditions for microbiological treatment will occur when the organic waste is nutritionally balanced as a fermentation medium: this will necessitate a pH favourable to bacterial growth (usually above pH 6·5), a plentiful supply of oxygen and suitable ratios of carbon to nitrogen and of carbon to phosphate. Microbiological treatment of waste results in incorporation of some of the soluble material into cellular substances that can then be removed as sludge, with simultaneous oxidation of the remainder to liberate carbon dioxide to the atmosphere and form water (see Chapter 12) that can be discharged by surface drainage. The separated sludge retains a capacity for endogenous oxidation of the reserve food materials of the cells, but if a brief period of self-digestion is permitted to the cells after their separation from the waste liquor, the sludge can then be discarded as harmless matter having a low oxygen demand.

2.1. *The activated sludge process for waste disposal*

Sewage, the predominantly organic municipal waste from industrial and domestic premises, is a dilute aqueous solution of organic and inorganic substances containing suspended solids, together with a mixed population of micro-organisms derived from human faeces and infiltration water. Microbiological degradation of urea to ammonia and carbon dioxide is usually almost completed before the sewage arrives at the treatment plant.

The ratio of organic matter to nitrogen and to phosphate in the waste largely determines whether it is amenable to biological treatment. B.O.D.:phosphorus, and B.O.D.:nitrogen ratios of 150:1 and up to about 30:1, respectively, have been recommended for treatment by the activated sludge process.

Treatment of sewage begins when floating matter and coarse suspended solids are screened from the waste on entry to the treatment plant. The separated solids are pulverized before re-entry to the main sewage stream, which now passes through a grit chamber to permit sedimentation of sand, grit and heavy mineral solids. The fine suspension of solids and soluble matter then enters the primary settlement tank for a brief period, to permit deposition of heavier particles as sludge. The "raw sludge" periodically removed from the sedimentation tank readily undergoes putrefaction by anaerobic microbes, and it is therefore treated separately.

The supernatant fluid now enters the aeration tank, where its nutrient status is adjusted, if necessary, by adding nitrogen and phosphate or other required nutrient before mixture with an inoculum of activated sludge. This inoculum is a floc comprising the cells of a mixed population of micro-organisms, predominantly zoogleal bacteria, which have been grown under aerobic conditions: it is used to initiate the main fermentation under aerobic conditions, preferably assisted by mechanical agitation and forced aeration. The B.O.D. of the waste entering the aeration tank must be limited to a level compatible with the rate of supply of dissolved oxygen. Suspensions with B.O.D. values of up to 3000–5000 mg/l. can frequently be employed in this process, but rapid mixing of the input material throughout the aeration tank is helpful in maintaining uniform fermentation conditions. Rapid growth of the microbial population ensues, with concomitant removal of soluble and insoluble organic matter by oxidation, by incorporation into structural and storage materials of the cells and by adsorption on the floc. Extension of the period of aeration is usually accompanied by respiration of cellular substrates, so that autolysis begins, with a resultant diminution of the microbial population. Froth formation, mainly related to proteinaceous materials in the waste, can be troublesome at this stage.

The effluent enters a secondary settlement tank, where sedimentation of stabilized sludge occurs. The clear liquid leaving the secondary settlement tank is then either chlorinated or re-aerated before disposal as a liquid of low B.O.D. to surface waters.

2.2. *Sludge digestion*

The stabilized sludge, containing 1–4% dry solids of high protein content, is partially recirculated as inoculum to the aeration tank of the activated sludge process, but the excess is mixed with the raw sludge from the primary settlement tank before disposal by incineration or composting or, alternatively, by anaerobic digestion in a sludge digester maintained at an elevated temperature favourable to bacterial metabolism.

Sludge digestion occurs in two stages. Heterotrophic bacteria growing on the complex organic substrate first form mixed organic acids, and these in turn are fermented by non-motile, non-sporing, gram-negative species of obligately anaerobic methane bacteria. The methane bacteria, as a physiological group, are limited in their oxidizable substrates to the lower fatty acids, the lower aliphatic alcohols and the gases hydrogen, carbon monoxide and carbon dioxide, and possibly also to acetone, 2,3-butylene-glycol and a few simple organic compounds: individual species exhibit extreme substrate specificity.

The products of anaerobic fermentation by methane bacteria are independent of the substrate utilized and consist simply of a mixture of methane and carbon dioxide, but details of the biochemical processes involved are not fully understood.

The methane gas produced during sludge digestion is used for production of heat and the digested sludge is dried for use as a fertilizer.

Suggested literature for further reading

1. BEHRMAN, E. J., Tryptophan metabolism in *Pseudomonas*, *Nature, Lond.* **196**, 150–2 (1962).

2. CHAPMAN, P. J. and DAGLEY, S., Oxidation of homogentisic acid by cell-free extracts of a *Vibrio*, *J. gen. Microbiol.* **28,** 251–6 (1962).
3. DAGLEY, S., CHAPMAN, P. J., GIBSON, D. T. and WOOD, J. M., Degradation of the benzene nucleus by bacteria, *Nature, Lond.* **202,** 775–8 (1964).
4. DAVIES, J. I. and EVANS, W. C., Ring fission of the naphthalene nucleus by certain soil Pseudomonads, *Biochem. J.* **85,** 21P (1962).
5. DAVIES, J. I. and EVANS, W. C., Oxidative metabolism of naphthalene by soil Pseudomonads, The ring-fission mechanism, *Biochem. J.* **91,** 251–61 (1964).
6. ELSDON, S. R. and PEEL, J. L., Metabolism of carbohydrates and related compounds, *A. Rev. Microbiol.* **12,** 145–202 (1958).
7. HAYAISHI, O., TANIUCHI, H., TASHIRO, M. and KUNO, S., Studies on the metabolism of kynurenic acid. I. The formation of L-glutamic acid, D- and L-alanine, and acetic acid from kynurenic acid by *Pseudomonas* extracts, *J. biol. Chem.* **236,** 2492–7 (1961).
8. HORIBATA, K., TANIUCHI, H., TASHIRO, M., KUNO, S. and HAYAISHI, O., The metabolism of kynurenic acid. II. Tracer experiments on the mechanism of kynurenic acid degradation and glutamic acid synthesis by *Pseudomonas* extracts, *J. biol. Chem.* **236,** 2991–5 (1961).
9. JENKINS, S. H., The composition of sewage and its potential use as a source of industrial water, *Chemy. Ind.* 2072–9 (1962).
10. JOHNSON, M. J., Utilization of hydrocarbons by micro-organisms, *Chemy. Ind.* 1532–7 (1964).
11. NISHIZUKA, Y., ICHIYAMA, A., NAKAMURA, S. and HAYAISHI, O., A new metabolic pathway of catechol, *J. biol. Chem.* **237,** PC 268–70 (1962).
12. ORNSTON, L. N. and STANIER, R. G., Mechanism of β-ketoadipate formation by bacteria, *Nature, Lond.* **204,** 1279–83 (1964).
13. PARKER, H. A., *Biological Formation of Methane*, Ciba lectures in microbial biochemistry: *Bacterial Fermentations*, pp. 1–27, Wiley & Sons, New York, 1956.
14. RIBBONS, D. W. and EVANS. W. C., Oxidative metabolism of protocatechuic acid by certain soil Pseudomonads: a new ring-fission mechanism, *Biochem. J.* **83,** 482–92 (1962).
15. ROGOFF, M. H., Oxidation of aromatic compounds by bacteria, *Adv. appl. Microbiol.* **3,** 193–221 (1961).
16. TRECCANI, V., Microbial degradation of hydrocarbons, *Prog. ind. Microbiol.* **4,** 1–33 (1963).
17. Symposium on fermentation waste disposal, *Biotechnology and Bioengineering* **4,** 127–211 (1962).
18. TANIUCHI, H. and HAYAISHI, O., Studies on the metabolism of kynurenic acid. III. Enzymatic formation of 7,8-dihydroxykynurenic acid from kynurenic acid, *J. biol. Chem.* **238,** 283–93 (1963).
19. TANIUCHI, H., HATANAKA, M., KUNO, S., HAYAISHI, O., NAKAJIMA, M. and KURIHARA, N., Enzymatic formation of catechol from anthranilic acid, *J. Biol. Chem.* **239,** 2204–11 (1964).
20. *Disposal of Industrial Waste Materials*, Society of Chemical Industry, 14 Belgrave Square, London.

CHAPTER 17

Industrial Uses for Micro-organisms in single step Enzymic Processes

THE industrial processes considered in earlier chapters have required a series of co-ordinated metabolic reactions for biosynthesis of the desired products or, alternatively, for oxidative degradation of aromatic compounds. By contrast, products from a number of other industrial microbiological processes may be considered to depend upon a single transformation in appropriate substrates: these are essentially one-step enzymic processes. Some of these reactions are of great antiquity (e.g. vinegar production), the process having been evolved on traditional lines of fermentation without knowledge of the underlying biochemistry. Others (e.g. transformation of steroids) are a direct result of purposeful research and are more akin to chemical reactions aided by versatile but labile catalysts in the form of microbial enzymes. Further developments in these techniques may provide the organic chemist with diverse microbiological catalysts capable of effecting specific, and sometimes novel, transformations by routes more direct than those of conventional chemical procedures. For the present, however, most microbiological transformations are of economic importance only when the product can command a high price and when the chemical process is less efficient.

1. Microbiological acetification and the production of vinegars by acetic acid bacteria

Microbiological acetification is essentially a one-step oxidative process characteristically performed by the acetic acid bacteria, a group of ellipsoidal to rod-shaped obligate aerobes included within the Order Pseudomonadales. The acetic acid bacteria are characteristically able to oxidize ethanol rapidly and completely to free acetic acid and to tolerate extremely acid conditions (2–11% of acetic acid in the medium). Two main genera may be distinguished, mainly by their ability to oxidize acetate.

(i) *Acetobacter*. This genus includes those organisms able to oxidize acetate and lactate to carbon dioxide and water. They may have peritrichous flagella or may be non-flagellated.

(ii) *Acetomonas* (*Gluconobacter*). Organisms of this genus are unable to oxidize acetate. They may have polar flagella or may be non-flagellated.

The genus *Acetomonas* includes the organisms useful for vinegar production but those belonging to the genus *Acetobacter* are recognized in industry as vinegar spoilage organisms.

Nutritionally, the acetic acid bacteria comprise a spectrum of nutritional types from the lactaphilic type (growing well on lactate as oxidizable substrate, but not growing on glucose media lacking lactate) at one end to the glycophilic type (growing well on glucose and on certain sugar alcohols, but not on lactate) at the other. The lactaphilic acetic acid bacteria have simple nutritional requirements, but the glycophilic organisms are more exacting, having requirements for certain preformed amino acids (e.g. glutamate, aspartate or proline) or for vitamins of the B complex.

The characteristic oxidation of ethanol to acetic acid performed by acetic acid bacteria is usually considered to proceed by two successive dehydrogenations that depend on the cytochrome system for transfer of electrons to atmospheric oxygen as the ultimate hydrogen acceptor:

$$\text{(a)}\quad \underset{\text{ethanol}}{CH_3CH_2OH} \xrightarrow{-2H} \underset{\text{acetaldehyde}}{CH_3CHO} + H_2O$$

$$\text{(b)}\quad CH_3CH(OH)_2 \xrightarrow[\text{(TPN-dependent dehydrogenase)}]{-2H} CH_3COOH + H_2O$$

hydrated form of acetaldehyde — acetic acid

It should be noted that acetic acid bacteria are able under anaerobic conditions to replace the second dehydrogenation, by a dismutation of acetaldehyde, whereby one molecule of substrate is oxidized to acetic acid at the expense of a second molecule, which is simultaneously reduced to ethanol. This is the reaction responsible for spoilage of wine by contaminating acetic acid bacteria:

$$CH_3CHO + CH_3CHO \longrightarrow CH_3COOH + CH_3CH_2OH$$

2 moles acetaldehyde — 1 mole acetic acid — 1 mole ethanol

Vinegars are produced by microbiological acetification and flavouring of a variety of alcoholic liquors that have been obtained by fermentation of raw materials containing starch or sugar: malt vinegar, wine vinegar, apple or cider vinegar are typical products, but a variety of other alcoholic fermented fruit juices are used on occasion. Direct acetification of dilute ethyl alcohol affords the spirit or white vinegar commonly used in the food industries.

The original and ancient process of acetification known as the Orleans or slow process was conducted in barrels aerated in each upper half by perforations in the sides. A small inoculum of naturally occurring acetic acid bacteria was provided by beginning with barrels one-third full of high-grade vinegar. Fresh wine was then added to provide the alcoholic substrate, and acetification proceeded slowly for about 5 weeks, with weekly additions of further substrate until the barrels were half-full. Withdrawal of the acetified product (vinegar) to reduce the retained volume to the original one-third with repetition of the serial additions of wine as substrate rendered the process semi-continuous but slow. The slow process yielded vinegar of high quality but it has now been largely replaced by more modern methods.

The modern quick vinegar process depends on providing a large surface for intimate contact between the bacteria, their substrate, and oxygen, so as to catalyse the oxidation of the ethanolic liquors to acetic acid. Alcoholic substrates, containing as little fermentable carbohydrate as possible, are percolated with re-circulation for several days through layers of birch twigs (besoms) contained in cylindrical wooden towers (generators). The besoms, by reason of their impregnation with old vinegar, provide an inoculum of acetic acid bacteria and an acidic environment that prevents undesirable bacterial contamination. The generators are cooled below 40°C to prevent loss of acetic acid by volatilization and to prevent inhibition by heat of the growth of the acetic acid bacteria. Excessive cooling must also be avoided as it slows down the acetification process and may permit development of *Acetobacter xylinum*, one of the encapsulated acetic acid bacteria that give an undesirable viscous or ropey consistency to the vinegar fluid because of their cellulosic capsules.

The freshly acetified vinegar is stored for about 6 weeks in wooden vats containing old beechwood chips to permit precipitation of colloidal materials together with iron, copper, calcium and phosphate. The finished vinegar has good colloidal stability and has lost much of its initial harsh flavour. It is diluted to 4–5% acidity (determined as acetic acid), coloured by caramel addition if necessary, filtered and sometimes pasteurized before transfer to casks or bottles for sale.

The quick vinegar process makes use of mixed cultures of acetic acid bacteria as inoculum but it seems probable that pure culture techniques may be adopted in time and that the process may eventually be superseded by the modern method of submerged, aerated culture in acetators, which has already been successfully used on a laboratory scale.

2. Acetic acid bacteria in relation to the industrial synthesis of L-ascorbic acid

The dehydrogenation of ethanol to acetic acid is only one of many similar oxidations that can be effected by acetic acid bacteria. Sometimes the substrate is completely oxidized to carbon dioxide and water but many strains cause only partial oxidation of the substrate without fragmentation of the carbon chain. Many polyols and other hydroxy compounds with secondary alcohol groups in suitable configuration may be readily dehydrogenated in this manner to the corresponding keto-compounds, so that the acetic acid bacteria are potentially valuable to organic chemists as stereospecific catalysts of single-step oxidations.

The industrial synthesis of L-ascorbic acid (vitamin C) for use as an antioxidant in food industries depends at one stage on the single-step microbiological oxidation of D-sorbitol to L-sorbose by acetic acid bacteria:

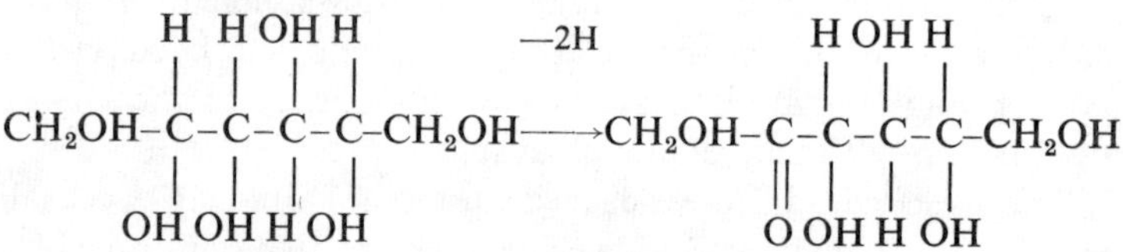

D-sorbitol L-sorbose

The synthesis of L-ascorbic acid begins with electrolytic reduction of D-glucose to D-sorbitol. Acetic acid bacteria (usually strains of *Acetobacter suboxydans*) are then used to dehydrogenate D-sorbitol to L-sorbose before the synthesis is completed by chemical stages involving oxidation of the diacetone derivative of L-sorbose to the carboxylic acid, removal of the acetone residues by acid hydrolysis and, finally, conversion of the methyl ester of 2-keto-L-gulonic acid to L-ascorbic acid.

The microbiological stage is conducted at 30–35°C in stainless steel aerobic fermenters containing the medium: 20% w/v sorbitol; 0·3% cornsteep liquor; accessory food factors as required by the particular strain employed. The fermentation medium is inoculated with 3% v/v of a culture of *A. suboxydans* in the logarithmic phase of growth in a similar medium. Further additions making a maximum of 30% sorbitol are made at intervals from the 24th hour and the oxidation is completed within 72–96hr when 90–95% of the D-sorbitol has been converted to L-sorbose.

3. Microbiological transformation of steroids

3.1. *Structure of steroids* (see Fig. 17.1)

Steroids are a group of related organic compounds structurally based on the tetracyclic skeleton of 1,2-cyclopentenophenanthrene (I). They may be conveniently considered as derivatives of cyclopentanoperhydrophenanthrene (II), an alicyclic saturated hydrocarbon designated sterane.

Substituents on the pairs of asymmetric carbon atoms at the junction of each pair of rings (*A*/*B*, *B*/*C*, *C*/*D*) may be either above

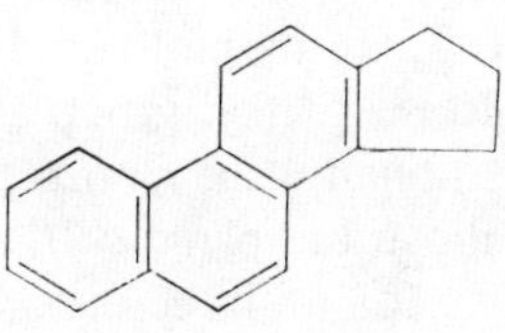

1, 2-cyclopenteno-
phenanthrene (I)

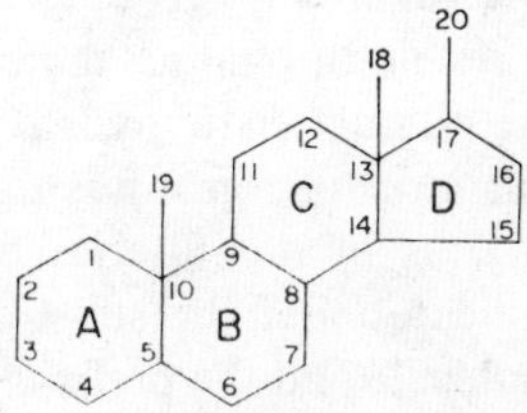

Cyclopentanoperhydro-
phenanthrene (II)
(sterane)

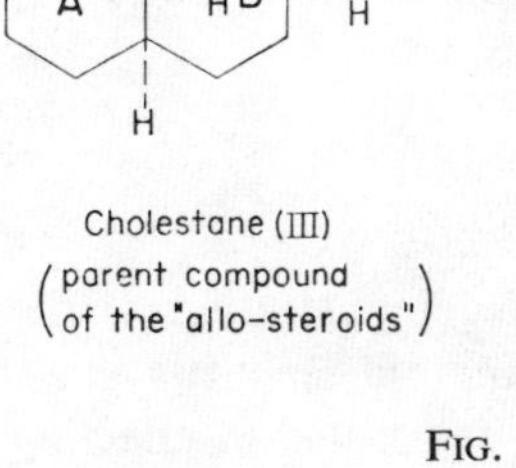

Cholestane (III)
(parent compound of the "allo-steroids")

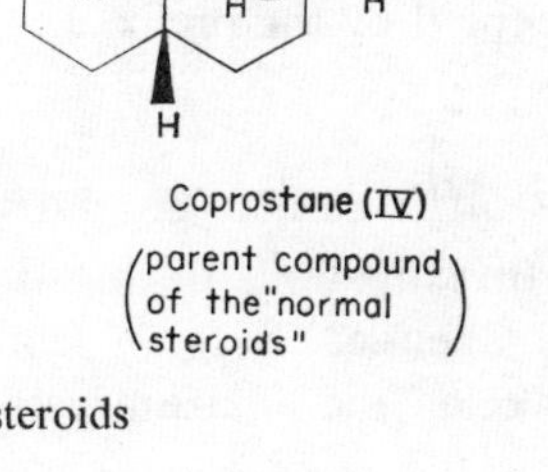

Coprostane (IV)
(parent compound of the "normal steroids")

FIG. 17.1 Skeleton of steroids

or below the plane of the ring: the compounds are said to be in the *cis*-form when both substituents are on the same side of the plane of the ring, but in the *trans*-form when they are on opposite sides. Steroid molecules are essentially flat, and this can result only when the manner of fusion of rings *B* and *C* gives the *trans*-form. Most of the naturally occurring steroids (sterols, bile acids, sex and adrenocortical hormones) have rings *B*/*C* and *C*/*D* fused to give the *trans*-forms, but they are differentiated into two series according to the form of fusion between rings *A* and *B*: the "cholestane series" (allo-steroids) (III) has the *trans*-form and the "coprostane series" (normal steroids) (IV) has the *cis*-form, as may be seen from Fig. 17.1. In both series the methyl groups at C_{10} and C_{13} are *cis*, and the side-chain at C_{17} is in *cis* relationship to these angular methyl groups.

An additional asymmetric centre is generated in the sterols by the characteristic hydroxyl substitution at C_3. In natural sterols this alcoholic group is in *cis* relationship to the methyl group at C_{10}, as in the β-series of sterols. Conventionally, the prefix β indicates that the substituent group lies above the plane of the molecule and the prefix α or epi denotes inversion to the epimer in which the substituent group lies below the plane of the molecule.

In summary, there are eight dissimilar asymmetric carbon atoms (3, 5, 8, 9, 10, 13, 14, 17) in the nucleus of a fully saturated sterol, and this makes possible a total of $2^8 = 256$ stereoisomerides. This number of possible isomers will be doubled for every additional asymmetric carbon atom present in the side-chain.

Finally, we must note that many of the steroidal hormones (e.g. cortisone, testosterone, progesterone) have a double bond between C_4 and C_5 and that this position of unsaturation is denoted by the prefix Δ_4.

3.2. *Microbiological transformation of steroids*

Investigations during the 15 years after demonstration of the therapeutic effect of cortisone in rheumatoid arthritis have revealed the amazing versatility of micro-organisms in effecting specific modifications of the steroid structure, often in a manner inaccessible to conventional chemical processes. This knowledge has made available various novel intermediates for chemical synthesis of new steroids, and many of these have proved to be of value in medicine.

The reactions effected by micro-organisms may be either oxidative (e.g. hydroxylation, epoxidation or dehydrogenation) or reductive (e.g. hydrogenation of a double bond or a carbonyl group). The key to a specific transformation usually resides in utilizing a particular strain having the desired enzymic constitution, in conjunction with a suitable substrate.

3.2a. *Hydroxylation.* Micro-organisms are able to introduce hydroxyl groups at specific positions in the steroid molecule. Mono-hydroxylations are known for positions 1α, 2β, 6β, 7α, 7β, 9α, 11α, 11β, 12β, 14α, 15α, 15β, 16α, 17α and for position 21 in the side-chain. Di-hydroxylations are known to occur in positions $6\beta + 11\alpha$, $6\beta + 14\alpha$, $7\alpha + 14\alpha$, $9\alpha + 14\alpha$, $11\alpha + 17\alpha$, $11\alpha + 21$, $11\beta + 21$, and $12\beta + 15\beta$.

Hydroxylation of the 11α-position is probably the most important of these for industrial purposes. A microbiological process converts progesterone (Δ^4, 3-keto-steroid), a commercially avail-

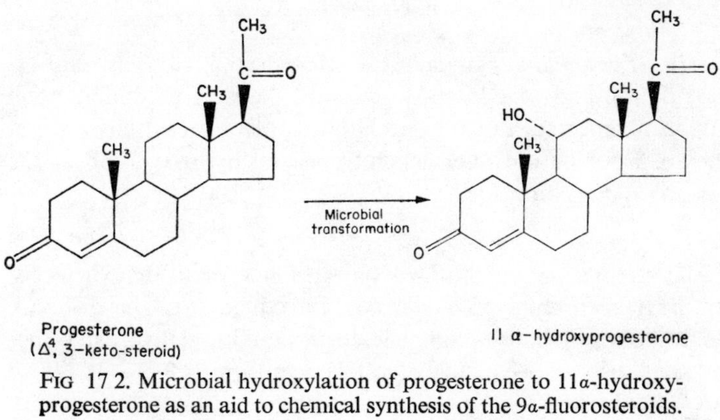

FIG 17 2. Microbial hydroxylation of progesterone to 11α-hydroxyprogesterone as an aid to chemical synthesis of the 9α-fluorosteroids.

able substrate prepared from the plant steroids, diosgenin and stigmasterol, to 11α-hydroxyprogesterone (Fig. 17.2). Chemical processes can then complete the synthesis of the 9α-fluoro steroids, the most active group of anti-inflammatory steroids.

Also of commercial importance is 16α-hydroxylation, because introduction of the 16α-hydroxyl group into the 9α-fluoro steroids removes the undesirable property causing high salt retention. A microbiological process for 16α-hydroxylation of 9α-fluorocortisol has been reported; the resultant 16α-hydroxy-9α-fluorocortisol (Fig. 17.3) can be converted by a chemical reaction to triamcinolone, a corticosteroid of medicinal importance.

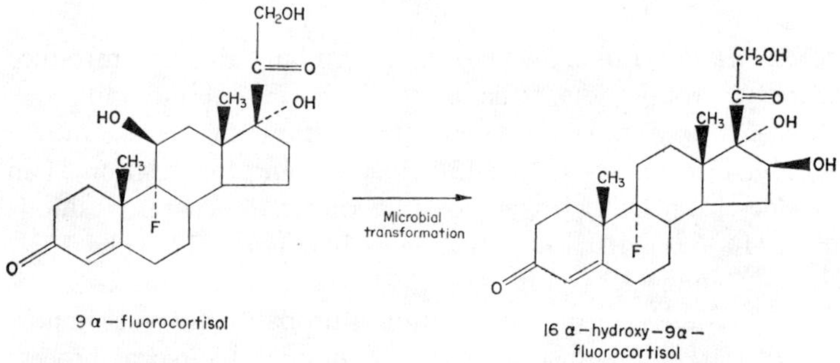

FIG. 17.3. Microbial hydroxylation of 9α-fluorocortisol to 16α-hydroxy-9α-fluorocortisol as an aid to chemical synthesis of triamcinolone.

3.2b. *Epoxidation.* In suitable unsaturated steroids 9β, 11-epoxidation and 14α, 15-epoxidation of the isolated double bond has been reported, but the mechanisms of the reactions are not known. Epoxidation may depend on axial hydroxylation at the position of unsaturation.

3.2c. *Dehydrogenation.* Two types of microbial dehydrogenation have been reported in steroids: introduction of the Δ^1-bond, the Δ^4-bond, or both, giving the corresponding dehydrosteroids; and oxidation of secondary alcohols at positions 3, 7, 11, 12, 17 and 20 to ketones.

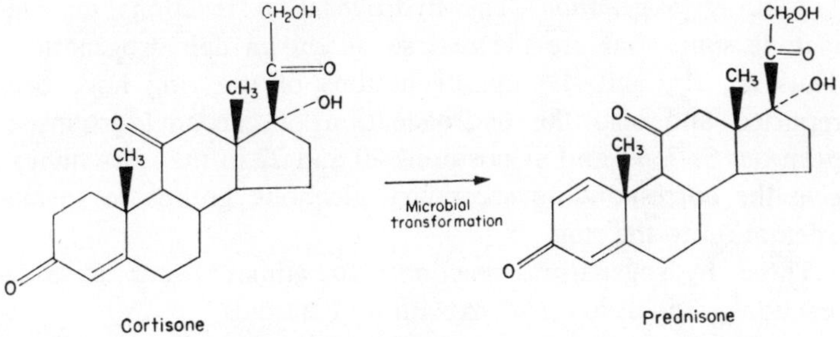

FIG. 17.4. Microbial dehydrogenation of cortisone as an aid to chemical synthesis of the 1-dehydrosteroids.

Conversion of cortisone to prednisone (Fig. 17.4), a compound with increased anti-inflammatory action, is typical of the dehydrogenation reaction that introduces the 1,2-double bond; and the formation of the hormone, testosterone, from 5-androstene-3β, 17β-diol illustrates the transformation of a secondary alcoholic group (see Fig. 17.5).

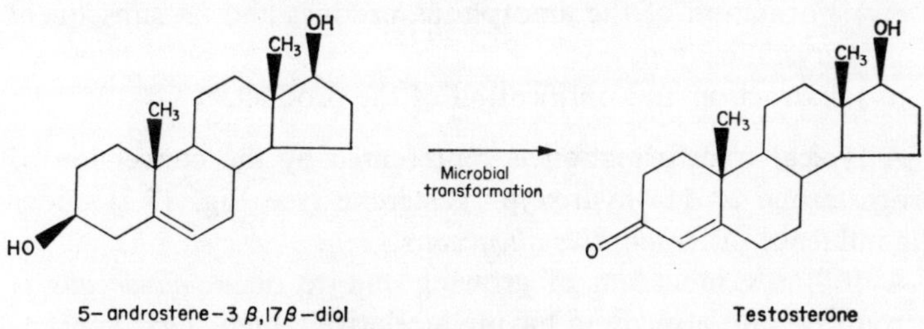

FIG. 17.5. Microbial dehydrogenation of a secondary alcoholic group to give a ketosteroid.

Some of these dehydrogenases appear to require NAD as hydrogen acceptor but in others the coenzyme is probably a flavoprotein.

3.2d. *Hydrogenation.* The hydrogenation reactions reported include some that are the reverse of known dehydrogenations. Thus Δ^1, Δ^4, and Δ^{16} hydrogenations of the ring have been reported and also the hydrogenation of carbonyl groups at positions 3, 7, 17 and at positions 20 and 22 in the side-chain to give the corresponding secondary alcoholic groups in specific orientation to the ring.

These hydrogenation reactions sometimes occur as side-reactions in microbial hydroxylation of steroids.

3.3. *Procedures for microbiological transformation of steroids*

These essentially enzymic processes may be considered in several stages.

(i) Production of a vegetative or sporulated culture of the required micro-organism.
(ii) Addition to the reaction mixture of a suitable substrate steroid and its uniform dissemination in the fluid.
(iii) Provision of optimum conditions for enzymic conversion of substrate to product.
(iv) Formation of the amorphous product and its subsequent crystallization.
(v) Extraction and purification of the product.

A typical transformation is represented by the conversion of progesterone to 11α-hydroxyprogesterone (see Fig. 17.2) under the influence of *Aspergillus olivaceous.*

A 10% v/v inoculum of growing culture of *A. olivaceous* is introduced into a medium having a relatively high sugar concentration (about 5%), supplied as glucose, sucrose, molasses or other readily utilizable sugar, a relatively low concentration of nitrogen (0·1–0·15% N), supplied as $NaNO_3$, $(NH_4)_2HPO_4$ or in organic form (e.g. cornsteep liquor, molasses) that has been adjusted to pH 6·5 and sterilized.

After inoculation, the culture is agitated and aerated to a relatively high degree for 24 hr, while being maintained at 28°C the optimum temperature for growth of this particular organism. Thick growth of mycelium occurs during the period accompanied by a fall in pH to 5·0, but only about half the sugar is consumed.

Progesterone is almost insoluble in water; it is therefore dissolved in a suitable carrier (e.g. propylene glycol) before addition at about 0·1% w/v concentration to the fermentation broth. The reaction converting progesterone to 11α-hydroxyprogesterone is completed within about 18 hr if good aeration and agitation are maintained. Alternatively, the progesterone substrate may be added semicontinuously in 0·015% w/v portions over a period of 24 hr, the transformation reaction being completed within about 4 hr after the last addition.

Division of the substrate into small portions added serially reduces the toxic effect of the solvent used for solution of the steroid substrate and may also increase the efficiency of conversion by minimizing loss due to side-reactions.

Extraction of the transformed product may present economic difficulties if the fermentation medium contains a large amount of anti-foam or if the growth of the micro-organism has caused other metabolic products to accumulate. Spores rather than vegetative mycelium have been used in some processes in an endeavour to overcome these difficulties.

A typical process involves using conidia of *A. ochraceous* produced by incubation at 28°C for 4–5 days on barley grains or on Sabouraud's dextrose agar in standard surface culture bottles. A suspension of spores is prepared by transfer to sterile 1% w/v phosphate buffer (pH 7·0) containing 0·01% v/v sodium monolauryl sulphonate or other suitable surface-active agent. Fragments of mycelium and other debris are removed by filtration through sterile glass-wool, the spores are washed by re-suspension in sterile buffer, washed in distilled water and finally re-suspended in phosphate buffer to give a suspension containing about 2×10^8 conidia/ml. Glucose 0·25% w/v, as energy source, and 0·5–0·1% progesterone (added as a 2·5% solution in propylene glycol, acetone or ethanol), as substrate, are added with agitation. The reaction mixture is then agitated and aerated at 24–33°C for 16–24 hr. Approximately 55% of the progesterone is converted to 11α-hydroxyprogesterone, and about 31% is transformed to 6β,11α-dihydroxyprogesterone.

The pH of the reaction does not appear to be critical within the range pH 4–8·5, phosphate does not appear to be essential, and the reaction is not affected by the inorganic ions from $FeSO_4$, $ZnSO_4$, $CuSO_4$, $MnSO_4$, $MgSO_4$, $Co(NO_3)_2$, by E.D.T.A. ($5–10 \times 10^{-4}$M), by ascorbic acid (10^{-2}M), cysteine (10^{-2}M), sodium fluoride (10^{-2}M), arsenate (10^{-3}M) or dinitrophenol (5×10^{-3}M). The transformation is, however, inhibited by sodium azide, mercuric chloride, chloro-mercuribenzoate and KCN, denoting the participation of SH-enzymes and the cytochrome system.

When the reaction is completed, the spores that remain ungerminated in the nitrogen-deficient medium may be separated by centrifugation and used as catalytic agents for further transformations with little loss of activity. The transformed product may then be extracted from the supernatant with a suitable solvent.

4. Production of (—)-ephedrine

(—)-Ephedrine or 1-phenyl-2-methyl-aminopropan-1-ol, an alkaloid having a physiological action similar to that of adrenaline, is obtained from plants of the genus *Ephedra* and has the structure:

$$C_6H_5.{}^*CHOH.{}^*\underset{\displaystyle NH.CH_3}{\underset{|}{CH}}.CH_3$$

The two dissimilar asymmetric carbon atoms (*) offer the possibility of four stereoisomers, but the (—)-form has much the greatest pharmacological activity.

A microbiological reaction for production of the specific isomer, (—)-ephedrine, was described in 1921. A strain of yeast is employed to produce acetaldehyde from glucose under suitable reducing conditions (see Chapter 10). Benzaldehyde is added to the fermentation broth, and a carboligase then couples the benzaldehyde and acetaldehyde substrates to form the optically active compound, 1-phenyl-1-hydroxypropan-2-one, in which the hydroxyl group has the same configuration as in (—)-ephedrine:

$$C_6H_5CHO + CH_3CHO \longrightarrow C_6H_5.\underset{\displaystyle \mathrm{OH}}{\underset{|}{CH}} - \overset{\displaystyle \mathrm{O}}{\overset{\|}{C}}.CH_3$$

The intermediate 1-phenyl-1-hydroxy-propan-2-one is then reduced with hydrogen under pressure in the presence of colloidal platinum and methylamine to give (—)-ephedrine.

Suggested literature for further reading

1. Allgeier, R. J. and Hildebrant, F. M., newer developments in vinegar manufacture, *Adv. appl. Microbiol.* **2,** 163–82 (1960).
2. Cheldelin, V. H., The acetic acid bacteria (Chapter 1 in *Metabolic Pathways in Micro-organisms*), 1–29, E. R. Squibb Lectures on Chemistry of Microbial Products, Wiley & Sons, New York and London, 1961.
3. Christner, J. E., Schlesinger, M. J. and Coon, M. J., Enzymatic activation of biotin: biotinyl adenylate formation, *J. biol. Chem.* **239** (11) 3997–4005 (1964).
4. DeLey, J. and Kersters, K., Oxidation of aliphatic glycols by acetic acid bacteria, *Bact. Rev.* **28,** 164–80 (1964).
5. El-Tayeb, O., Knight, S. G. and Sih, C. J., Substrate specificity in steroid side-chain degradation by micro-organisms, *Biochim. biophys. Acta* **93,** 411–7 (1964).
6. El-Tayeb, O., Knight, S. G. and Sih, C. J., Steroid epoxide cleavage by *Cylindrocarpon radicicola*, *Biochim. biophys. Acta* **93,** 402–10 (1964).
7. Karow, E. O. and Petsiavas, D. N., Effect of physical variables on microbiological transformation of steroids, *Ind. Engng. Chem. ind.* (*int.*) *Edn.* **48,** 2213–17 (1956).
8. Neuberg, C. and Hirch, J., Uber ein Kohlenstoffketten knupfendes Ferment (Carboligase), *Biochem. Z.* **115,** 282–310 (1921).
9. Prescott, S. C. and Dunn, C. G., The Microbiological transformation of steroids (Chapter 44 in *Industrial Microbiology*), McGraw-Hill, New York, Toronto and London, 1959.
10. Rainbow, C., The biochemistry of the Acetobacters, *Prog. ind. Microbiol.* **3,** 43–70 (1961).
11. Rao, M. R. R., Acetic acid bacteria, *A. Rev. Microbiol.* **11,** 317–38 (1957).
12. Stodola, F. H., *Chemical Transformations by Micro-organisms*, Chapman & Hall, London, 1958.

13. Stoudt, T. H., The microbiological transformation of steroids, *Adv. appl. Microbiol.* **2,** 183–222 (1960).
14. Tamm, C. H., Conversion of natural substances by microbial enzymes, *Angew. Chem.* **1,** 178–95 (1962).
15. Titus, E., The metabolism of cardiac lactones by micro-organisms, *Adv. appl. Microbiol.* **3,** 279–92 (1961).
16. Tsong, Y. Y., Wang, K. C. and Sih, C. J., Mechanisms of steroid oxidation by micro-organisms. V. Reduction of the $\triangle^{6,7}$ double bond, *Biochim. biophys. Acta* **93,** 398–401 (1964).
17. Végina, C., Sehgal, S. M. and Singh, K., Transformation of steroids by spores of micro-organisms. I. Hydroxylation of progesterone by conidia of *Aspergillus ochraceus, Appl. Microbiol.* **2,** 50–7 (1963).
18. Wallen, L. L., Stodola, F. H. and Jackson, R. W., *Type Reactions in Fermentation Chemistry*, Agricultural Research Service, U.S.A. Department of Agriculture, ARS-71-13, May 1959.

CHAPTER 18

Some Screening Methods Employed in the Search for New Antibiotics

THE discovery of new compounds with desirable biological activity depends on meticulous choice of techniques for testing the input materials and on weighing the statistical probability in favour of a successful outcome. When other factors are equal, the advantage will clearly lie in favour of operation on a large scale. Compounds "screened" in the chemical industry are usually of known structure, synthesized and partially purified by organic chemists: industries based on microbiological processes more usually investigate various compounds of unknown structure, in a crude state, produced by the physiological activities of micro-organisms (e.g. fungi, actinomycetes, bacteria).

The soil is a convenient and vast reservoir of diverse micro-organisms potentially able to provide input materials for the microbiological studies used in the search for new antibiotics, but other sources (e.g. plant pathogenic fungi, thermophilic actinomycetes) are also worth consideration.

It is essential to have a clear conception of objectives before beginning to screen, to adhere closely to these "targets" until they are superseded by competitive commercial developments, and to do everything possible to minimize time wasted on substances that eventually prove to be known antibiotics.

1. Primary *in vitro* screen

Soil micro-organisms may be isolated by inoculating serial

dilutions, from a suspension of soil in water, into agar media able to support growth of the desired isolates. For actinomycetes, in particular, media containing different carbon sources may help to differentiate between the isolated organisms. Incubation for about 7 days permits development of small discrete colonies that may be transferred to agar slopes of suitable maintenance media contained in small surface culture bottles. The principles of maintenance of these reference cultures have been described in Chapter 4.

Most methods for initial demonstration of antibiosis depend upon determining the inhibition zones produced when a well-grown culture of the isolated strain is placed on a microbiological assay plate seeded with the test pathogens that need to be controlled. Plugs of 10 mm diameter removed from well-grown surface cultures, sometimes from several different media, may be conveniently used for this purpose. Alternative methods have been devised to permit demonstration of antibiosis in the original colonies isolated from soil (e.g. the replica-plate method, whereby the original well-isolated colonies are transferred by means of a sterile velveteen stamp to four or more replica plates, of which each can be "seeded" with a different target organism).

Strains producing inhibition zones in surface cultures are then tested for antibiotic production in shake-flask submerged cultures. At this stage it is usual to employ a number of different media in order to direct the metabolism of the microbial culture along different pathways; the broths are assayed against the target organisms at intervals of 24 hr, usually between the second and fourth days. Media of several general types have proved useful.

(i) Medium with carbon source mainly derived from oil.
(ii) High carbohydrate, low phosphate medium with readily available nitrogen.
(iii) Medium having a high concentration of slowly utilizable carbohydrate, together with plentiful phosphate and limited nitrogen.
(iv) Czapek's sucrose–nitrate medium.

2. Secondary screen

Antibiotics giving adequate zones of inhibition (e.g. exceeding 15–20 mm diameter) against the chosen target pathogens are again produced in broth consisting of the best of the submerged fermentation media. Methods for extracting the active ingredient are determined by subjecting portions of broth at the time of peak titre to a variety of solvent extraction procedures over a wide range of pH, with cup-plate assay of extracts and spent broths: stability of the antibiotic to acid and alkali and to elevated temperature is usually determined at the same time. Adsorption or non-adsorption of the antibiotic on ion exchange resins is determined in a similar manner by assay of the resin filtrates and eluates. A one- or two-stage extraction procedure suitable for preparing a crude concentrate of the antibiotic may usually be devised from these results.

The crude antibiotic concentrates are scrutinized to eliminate, so far as possible, known antibiotics. Ultraviolet absorption spectrum, chromatograms and bio-autographs, extraction pattern, toxicity, spectrum of cross-resistance with known antibiotics and spectrum of biological activity are some of the properties that have proved useful for this purpose.

Further test of this crude antibiotic, after removal of solvent from the extract, will depend on the objectives of the examination and on the relative weight attributed to specific properties. Formulation is seldom attempted at this stage, but whenever possible tests are conducted on solutions (aqueous or in 50% acetone or in 7–50% *N*,*N*′-dimethylformamide) or on fine particle suspensions.

For antibiotics having potential use in medicine it is usual to assay the antibiotic by the cup-plate method in media containing 50% v/v whole blood; zones of inhibition diminished by more than half under these conditions are generally taken as indicating elimination of antibiotics inactivated by blood. Similar assays in media containing 10–50% v/v serum are used to recognize antibiotics that are bound in high degree by protein. Acid stability (especially desirable for an oral antibiotic), resistance to enzymes of various animal organs (e.g. liver) and sometimes resistance to penicillinase may be used as additional means of assessment at this stage.

Degree of mammalian toxicity is of prime importance in an antibiotic having a potential use in medicine. It is therefore usual practice to determine on aqueous concentrates of the crude antibiotic (conveniently produced in 5 l. mechanically agitated and aerated laboratory fermenters), the relationship between intraperitoneal toxicity in mice and the dose conferring protection when administered by a recognized route (e.g. subcutaneously for protection tests against virulent staphylococci). Antibiotics are rejected if they fail to give an adequate, even though arbitrarily specified, margin between the toxic and protective doses.

Antibiotics of potential use in agriculture, on the other hand, are examined differently. Stability to homogenates of plant tissues, stability to heat, exceptionally high biological activity, desired spectrum of biological activity and ability to move through discs of plant tissues (local systemic action) are some of the attributes that may be deemed desirable. Phytotoxicity of the crude antibiotic and its ability to protect plants against the target pathogens is usually determined in greenhouse tests before selecting those antibiotics worthy of more detailed evaluation.

New antibiotics of potential use in medicine that have survived thus far must be produced on a larger scale (usually in pilot-plant fermenters) to permit purification before being subjected to detailed pharmacological evaluation. Eventually, the few may arrive at the stage when consideration can be given to clinical evaluation and the formulations that may be suitable for their presentation.

Antibiotics of potential use in agriculture must also be produced on a relatively large scale to permit pharmacological testing and evaluation in greenhouse and small-plot trials, but these procedures differ little from the methods employed for evaluation of synthetic pesticides. A number of years may still be needed for field trials and commercial evaluation of the formulated antibiotic and for improvements in fermentation yield and in methods of extraction necessary to give an economically viable production process.

Suggested literature for further reading

1. BENTLEY, R., Biochemistry of Fungi, *A. Rev. Biochem.* **31,** 589–624 (1962).
2. BU'LOCK, J. D., Intermediary metabolism and antibiotic synthesis, *Adv. appl. Microbiol.* **3,** 293–342 (1961).
3. CAMPBELL, A. H., The search for new antibiotics, *Br. Med. Bull.* **16** (*Antibiotics in Medicine*), 82–85 (1960).
4. HARRIS, D. A. and RUGER, M. L., Microbiological aspects of new antibiotic screening. I. Plug test procedure, *Antibiotics Chemother*, **3,** 265–70 (1953).
5. LECHEVALIER, H. A. and CORKE, C. T., The replica plate method for screening antibiotic-producing organisms, *Appl. Microbiol.* **1,** 110–12 (1953).
6. MURAT, A., STINEBRING, W. R., SCHAFFNER, C. P. and LECHEVALIER, H., Screening for antibiotics active against intracellular bacteria, *Appl. Microbiol.* **7,** 109–12 (1959).
7. RHODES, A., FANTES, K. H., BOOTHROYD, B., MCGONAGLE, M. P. and CROSSE, R., Venturicidin: a new antifungal antibiotic of potential use in agriculture, *Nature, Lond.* **192,** 952–4 (1961).
8. SCHABEL, F. M. Jr. and PITTILLO, R. F., Screening for biological and characterization of antitumour agents using micro-organisms, *Adv. appl. Microbiol.* **3,** 223–56 (1961).
9. SHIBATA, S., NATORI, S. and UDAGAWA, S., *List of Fungal Products*, National Institute of Hygienic Sciences, Tokyo, 1961.
10. WAKSMAN, S. A., *Antibiotics of actinomycetes. The Actinomycetes* III, Baillière, Tindall & Cox, London, 1962.

CHAPTER 19

Continuous Culture of Micro-organisms

GROWTH of a micro-organism during a batch fermentation conforms to the characteristic growth curve described in Chapter 8. The lag phase or period of adaptation to environment leads to the logarithmic phase, in which the organism divides exponentially. This, in turn, is terminated by progressive decrements in the rate of growth, due to limitation of one or more of the essential nutrients, until the stationary phase is reached when the amount of protoplasm in the culture remains constant.

Evidently, time spent in the lag and stationary phases is wasteful for production of microbial cells, and for this reason commercial yeast manufacture (Chapter 8) and the brewing of beer (Chapter 10) are being gradually converted to newly developed methods of continuous fermentation.

The simplest form of apparatus needed for continuous culture of micro-organisms is illustrated diagrammatically in Fig. 19.1.

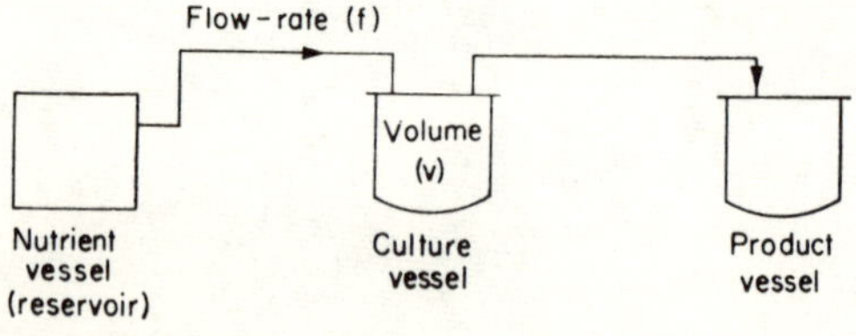

FIG. 19.1. Continuous culture.

The organism is grown in the culture vessel under the desired environmental conditions, fresh culture medium being continuously supplied at a constant rate (f). A constant-level overflow permits fermented liquor to be withdrawn into the product vessel at a rate equal to that of the medium input.

1. Monod's theory (extended by Herbert)

In 1950 Jacques Monod advanced a theory of continuous culture based upon two simple assumptions.

(1) Micro-organisms grow in an exponential manner with a maximum specific growth rate μ_m until the rate of cell division is limited by some environmental condition, giving a specific growth rate μ.
(2) The specific growth rate (μ) of any given microbial culture is proportional to the limiting substrate concentration (S), but approaches a maximum value μ_m which is reached when the substrate concentration is no longer limiting.

Putting the two assumptions into mathematical terms, we have

(1) $$\frac{dx}{dt} = \mu x.$$

Therefore

$$\mu = \frac{1}{x} \cdot \frac{dx}{dt} = \frac{d(\log_e x)}{dt} \qquad (19.1)$$

where x is the concentration of organism (mg dry weight/ml), and μ is the exponential or specific growth rate.

Equation (19.1) relates the specific growth rate (μ) to a function of the rate of change in concentration of the organism. Therefore, in the special case when the concentration of the organism is doubled in a time interval t_d, we can say

$$\mu = \frac{\log_e(2x/x)}{t_d} = \frac{\log_e 2}{t_d} = \frac{0{\cdot}693}{t_d}$$

(2) $$\mu = \mu_m \frac{S}{S + K_s} \qquad (19.2)$$

where K_s is a "saturation constant" numerically equal to the substrate concentration at which the growth rate is one-half the maximum value. (Compare the Briggs–Haldane or Michaelis equation in Chapter 6, which relates the rate of reaction (v) with substrate concentration (S) for any particular concentration of enzyme: $v = (V_L S/S + K_m)$.)

Experience has shown that the original assumptions embodied in eq. (19.2) are consistent with certain experimentally ascertained facts: (1) growth rate is nil if the limiting substrate concentration (S) is nil; (2) as S becomes large, so $\mu \longrightarrow \mu_m$; (3) if S is some small value, then $\mu \simeq \mu_m$ (S/K_s). In other words, $\mu \simeq$ constant $\times$ S, and the specific growth rate is therefore proportional to the limiting substrate concentration.

Considering Fig. 19.1, we may define "dilution rate" as $D = f/v$ and its reciprocal, the mean residence time, as $1/D = v/f$.

Then the rate at which cells leave the culture vessel is given by $-(\mathrm{d}x/\mathrm{d}t) + Dx$.

The net increase in growth ($\mathrm{d}x/\mathrm{d}t$) in the culture vessel must be the difference between the rate of growth (μx) and the rate of withdrawal:

$$\frac{\mathrm{d}x}{\mathrm{d}t} = \mu x - Dx = (\mu - D)x. \qquad (19.3)$$

In the steady state the specific growth rate (μ) must equal the dilution rate (D), because there is no change in cell concentration ($\mathrm{d}x/\mathrm{d}t = 0$). This will be true until the dilution rate exceeds a certain critical value (D_c), when the organism is washed out of the system.

1.1. *The relationship between the concentration of substrate and the concentration of cells in the steady state*

If it is also assumed that the rate of substrate consumed ($\mathrm{d}s/\mathrm{d}t$) is proportional to the rate of growth of cells, then

$$-\frac{\mathrm{d}s}{\mathrm{d}t} = \frac{\mathrm{d}x/\mathrm{d}t}{Y},$$

where Y (the Yield Constant) is the ratio of the weight of cells formed in finite time to the corresponding weight of substrate consumed.

Organisms in the culture vessel utilize substrate, thereby establishing a concentration gradient from S_R in the reservoir to S, the substrate concentration common to both the culture vessel and the product vessel. A mass balance for the substrate concentration in the culture vessel may now be constructed. The net rate of change of substrate concentration in the culture vessel ($\mathrm{d}S/\mathrm{d}t$) is given by the input (flow rate per unit volume $\times$ S_R) minus the output (flow rate per unit volume $\times$ S) minus the rate of substrate consumption by the growing cells:

$$\frac{\mathrm{d}S}{\mathrm{d}t} = \frac{f}{v}\cdot S_R - \frac{f}{v}\cdot S - \left(-\frac{\mathrm{d}s}{\mathrm{d}t}\right)$$

$$= DS_R - DS - \frac{\mathrm{d}x/\mathrm{d}t}{Y}$$

$$= DS_R - DS - \frac{\mu x}{Y} \qquad (19.4)$$

Hence, by substituting the value of μ obtained from eq. (19.2),

$$\frac{\mathrm{d}S}{\mathrm{d}t} = D(S_R - S) - \frac{\mu_m x}{Y}\left(\frac{S}{S + Ks}\right). \qquad (19.5)$$

Similarly, by substitution in eq. (19.3),

$$\frac{\mathrm{d}x}{\mathrm{d}t} = x\mu_m\left(\frac{S}{S + Ks}\right) - Dx. \qquad (19.6)$$

The two equations, (19.5) and (19.6), completely define the behaviour of unicellular organisms in continuous culture. In the steady state the concentration of cells and of limiting substrate in the culture vessel remain constant (i.e. dx/dt and dS/dt must be zero), so that eq. (19.5) becomes in the steady state

$$DS_R - D\tilde{S} = \frac{\mu_m \tilde{x}}{Y}\left(\frac{\tilde{S}}{\tilde{S} + Ks}\right), \tag{19.7}$$

where $\tilde{S}$ and $\tilde{x}$ are the symbols for the steady-state concentration of substrate and cells respectively.

Similarly, eq. (19.6) gives

$$\mu_m\left(\frac{\tilde{S}}{K_s + \tilde{S}}\right) = D. \tag{19.8}$$

From eq. (19.7)

$$S_R - \tilde{S} = \frac{\mu_m x}{Y}\left(\frac{\tilde{S}}{\tilde{S} + K_s}\right)/D.$$

Therefore, substituting the value of D obtained from (19.8),

$$S_R - \tilde{S} = \frac{\tilde{x}}{Y},$$

and by rearrangement of (19.8) we get

$$\tilde{S} = K_s\left(\frac{D}{\mu_m - D}\right). \tag{19.10}$$

Combining (19.9) with (19.10),

$$\begin{aligned}\tilde{x} &= Y(S_R - \tilde{S})\\ &= Y\left[S_R - K_s\left(\frac{D}{\mu_m - D}\right)\right].\end{aligned} \tag{19.11}$$

Herbert has plotted eq. (19.10) and (19.11) assuming $\mu_m =$ 1·0 hr^{-1}, $Y = 0{\cdot}5$, $K_s = 0{\cdot}2$ g/l. and $S_R = 10$ g/l. to give the curves shown in Fig. (19.2). The graph shows that, over a wide range of dilution rates, both the organism concentration ($\tilde{x}$) and the unused substrate concentration ($\tilde{S}$) remain nearly constant. The unused substrate concentration does not appreciably increase, and the organism concentration does not appreciably decrease

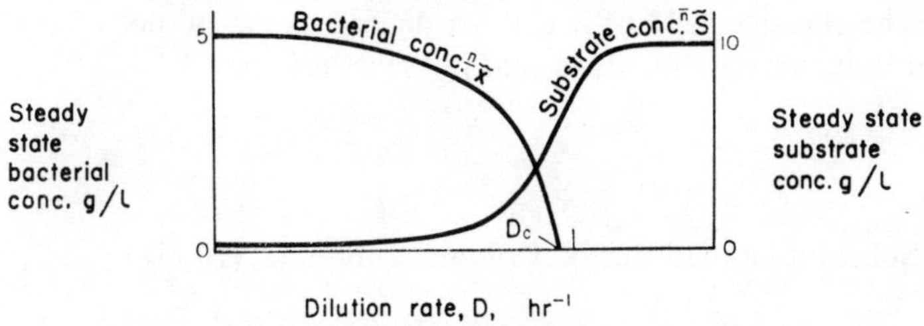

FIG. 19.2. Continuous culture: relationship between the bacterial concentration, the substrate concentration and the dilution rate.

until the dilution rate approaches the critical dilution rate D_c. In practice the concentration of the unused substrate is found to be low: therefore a continuous fermentation may be expected to give efficient utilization of substrate for growth over a wide range of dilution rates.

1.2. *Optimum conditions for the production of micro-organisms in a continuous fermentation*

Output of cells per unit volume of culture is equal to $D\tilde{x}$, and this value is obtained from eq. (19.11):

$$D\tilde{x} = DY\left[S_R - K_s\left(\frac{D}{\mu_m - D}\right)\right]. \qquad (19.12)$$

Equation (19.12) shows that the output for any particular substrate concentration in the reservoir is a function only of the properties of the organism (Y, K_s and μ_m) and the dilution rate (D). The output increases with increasing dilution rate up to a maximum dilution rate (D_m). The value of D_m can be found either graphically by plotting eq. (19.12) against values of D or, alternatively, by differentiating eq. (19.12) with respect to D and then equating to zero to give the expression

$$D_m = \mu_m\left[1 - \sqrt{\left(\frac{K_s}{K_s + S_R}\right)}\right]. \qquad (19.13)$$

The effective yield (Y_E) may be defined as the output of cells per unit substrate input, so that in the steady state

$$Y_E = \frac{\tilde{x}}{S_R}$$

Substituting the value of $\tilde{x}$ obtained from eq. (19.11),

$$Y_E = \frac{Y[S_R - K_s(D/\mu_m - D)]}{S_R} \tag{19.14}$$

The efficiency of utilizing limiting substrate (i.e. the ratio of the substrate used to the substrate input) at the maximum dilution rate (D_m) is Y_E/Y, and this value may be obtained from eq. (19.13) and (19.14):

$$\frac{Y_E}{Y} = \frac{(S_R/K_s + 1) - \sqrt{[(S_R/K_s + 1)]}.}{S_R/K_s} \tag{19.15}$$

Equation (19.15) shows that, at the maximum dilution rate, the maximum efficiency of utilization of limiting substrate depends only on S_R/K_s, and that maximum efficiency is obtained when the substrate concentration in the reservoir is as high as possible.

Thus far we see that a continuous fermentation designed for production of cell material has a maximum output at maximum dilution rate (D_m) given by eq. (19.13), and this rate will be just below the critical dilution rate (D_c) at which the organism would be washed out of the system: the substrate concentration in the reservoir for maximum efficiency should just exceed the limiting concentration.

2. Some examples of continuous fermentation

2.1. *Production of cellular material*

Continuous fermentation of micro-organisms, exemplified by the production of bakers' yeast, may be regarded as being both a stable and a self-regulating process.

Equation (19.10) shows that the substrate concentration within the fermenter when in the steady state ($\tilde{S}$) is a function only of the dilution rate (D): it is independent both of the concentration of the inflowing and limiting substrate (S_R) and of the cell concentration ($\tilde{x}$). Since K_s normally has a small numerical value it is evident that there will be little unused substrate ($\tilde{S}$) in the harvested broth.

Equation (19.11) shows that the concentration of cells in the fermenter ($\tilde{x}$) is a function of the substrate concentration in the reservoir (S_R). The value of the dilution rate D (i.e. flow rate/volume of fermenter), or the throughput, only becomes important when the value of the maximum specific growth rate (μ_m) for the particular organism is approached: at this time the expression ($D/\mu_m - D$) becomes large making the cell concentration in the steady state ($\tilde{x}$) low.

It should now be clear that a single-stage continuous fermentation for production of unicellular organisms is advantageous because it eliminates time wasted in "turn-round" and also ensures efficient use of substrate.

Attainment of the steady state in a continuous fermentation permits better adjustment for optimum medium constituents than is possible in a batch fermentation in which the physiology of the organism is changing continuously.

Experience shows that if the concentration of any substrate is high (say greater than 0·1M) then it is probable that the substrate is being wasted because the concentration exceeds the limiting value. When the concentration of a limiting substrate is increased the system will respond by increased cell growth, always provided that there is no other adverse effect on the environment.

Cellular material produced by continuous fermentation is sometimes used as food, and in these processes precautions must be taken to prevent the growth of potentially toxic mutant strains or of contaminating micro-organisms.

Any mutant or infecting organism likely to be a hazard must first compete with the established production strain for substrate and must also be immune to any indigenous antibiotic. Equation (19.11) shows that any foreign organism cannot predominate in the steady state unless it has a lower saturation constant (K_s) or a higher maximum specific growth rate (μ_m) than those of the established production strain. Thus in a continuous fermentation only a few contaminants will succeed. This contrasts with a batch fermentation in which a foreign organism having a higher saturation constant or a lower maximum specific growth rate may nevertheless predominate in the early stages if its lag phase is shorter than the similar, corresponding period required by the fermenting organism.

The Distillers Company Limited have reported that the maximum capacity of a fermentation plant producing bakers' yeast was increased by nearly 50% when converted from the batch process to continuous operation on a 7 day week. For practical reasons, six small fermenters (Fig. 19.3) connected in series were used in preference to one large fermenter. Cleaning was facilitated by

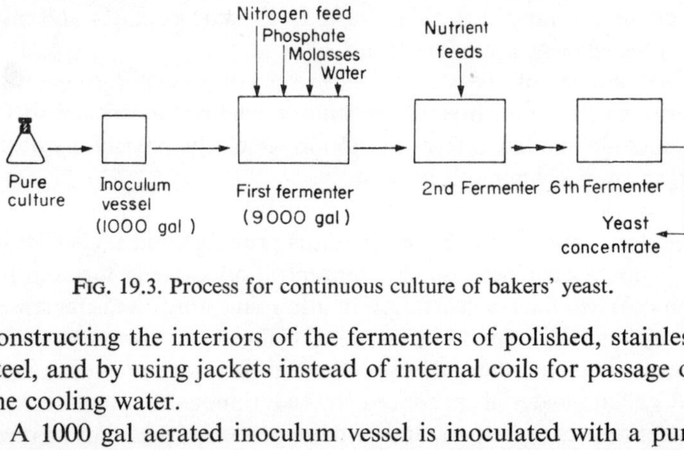

FIG. 19.3. Process for continuous culture of bakers' yeast.

constructing the interiors of the fermenters of polished, stainless steel, and by using jackets instead of internal coils for passage of the cooling water.

A 1000 gal aerated inoculum vessel is inoculated with a pure culture of yeast and after approximately 20 hr growth the entire volume is transferred to the first 9000 gal aerobic fermenter. Nutrients are supplied at a regulated rate: molasses to give maximum yield of yeast without production of alcohol; nitrogen (mixture of ammonium sulphate and ammoniacal liquor); and phosphate (diammonium hydrogen phosphate) and magnesium sulphate to give the correct quality of yeast. When the first fermenter is full the excess is pumped to the second fermenter, and then in sequence to the third, fourth and fifth aerated fermenters, all of which are fed with nitrogen, phosphate and molasses. The sixth and last fermenter in the chain is used to ripen the yeast by limitation of aeration and nutrient supply.

2.2. *Continuous disposal of waste products*

Many chemicals (e.g. phenols, thiosulphates, and ammonia from coke oven works) present difficulties in waste disposal because of their toxicity to micro-organisms, but most of these substances can be efficiently degraded in continuous fermentation processes. The steady-state concentration of the limiting substrate in a continuous fermentation process is low and this permits operation below the toxic concentration. Suitable organisms for these processes of waste disposal may be readily isolated from soil: pseudomonads oxidize phenols to carbon dioxide and water (see Chapter 16), thiosulphate is oxidized to sulphate by thiobacilli, and ammonia is oxidized by bacteria of the *Nitrosomonas* group.

Phenol, because of its inhibitory action on thiobacilli, and *Nitrosomonas* species, is removed by oxidation in the first stage of the process. The phenol aeration tanks may be constructed in concrete, and they must be suitably aerated and agitated. Sterility is of minor importance, and the tanks may therefore be open to the atmosphere. For efficient oxidation it is necessary to supply nitrogen (ammonium chloride), phosphoric acid, ferric chloride and magnesium sulphate and to maintain the pH between 7·5 and 8·0 at a temperature of 30°C.

The process begins in a medium containing only 100–350 μg/ml phenol, but later the phenolic feed may contain as much as 1500 μg/ml. The outflow from the phenol aeration tank passes to a settling tank, where it remains for an average period of 8 hr: the sludge that sediments is returned to the fermenter. The supernatant percolates through a coke filter-bed, which removes the organic material and permits aerobic micro-organisms to complete the detoxification of both phenol and thiosulphate. The degree to which thiosulphate oxidation occurs in the phenol aeration tank will depend upon the phenol concentration and the extent to which the thiobacilli have been inhibited.

Oxidation of ammonia occurs in the final stage: the plant design is similar to that employed in the first stage, but the effluent is now slightly acidic, so that it is necessary to use a metal non-toxic to the organism, such as cast iron, for construction of the tank.

2.3. *Continuous brewing*

Continuous brewing of beer was first reported at the beginning of this century. In this process, wort was passed by gravity through a series of vats. Practical difficulties in maintaining sterility and in control of taste (owing to proteinaceous precipitation) precluded its general use at that time.

Production of beer occurs in two stages (see Chapter 10): yeast cells are produced aerobically, but this growth stage is followed by an alcoholic fermentation stage which requires anaerobic conditions. Some of the yeast can be recovered from the anaerobic stage and recycled: this is the basis of the continuous brewing methods used in Canada and New Zealand.

Subsequent work has shown that wort can be continuously fermented in a single vessel if the concentration of oxygen is made the limiting substrate for growth of the yeast cells. The fermentation is maintained in the steady state to give a high concentration of yeast cells, so that the rate of fermentation is approximately three to four times greater than in a corresponding batch process. The fermentation can be further accelerated by using a higher temperature (above 20°C). Batch fermentations operated at elevated temperature would be marred by production of astringent flavours, but the continuous fermentation, now some twenty times faster than the batch fermentation, remains unaffected by reason of the much reduced contact time.

2.4. *Continuous production of vitamin B_{12}.*

The cost of manufacturing a fermentation product is a function of a number of variables, including the capital outlay in construction of the plant, the yield per unit volume (titre), the medium cost per unit titre, the extraction cost per unit titre and the rate of throughput.

Costs per unit working volume of construction, of medium, and of extraction are approximately the same for both batch and continuous fermentation processes. The relative economics of batch and continuous fermentations will therefore be largely determined by the extent to which the increased rate of throughput of the continuous process can offset any possible diminution in titre. When the product yield is high (e.g. in industrial alcohol and acetone fermentations) the influence of lowered titre on the cost of the continuous process will not be so serious as when the titre is relatively low (e.g. in antibiotic fermentations): the economics of the continuous process are therefore much more favourable for products that can be obtained in high yield.

Maximum titre in any fermentation process is achieved by keeping the maximum number of cells in the correct environment for as long as possible. Some complex products require prior formation of intermediates, so that a sequence of optimum conditions may be necessary for the overall biosynthesis. This is accomplished in a continuous fermentation when broth is permitted to flow slowly through a series of fermenters, each operating under conditions optimum for a particular step in the biosynthesis.

Vitamin B_{12} is an extreme example of a complex product obtained in low yield. The continuous process is feasible, but there must be no marked diminution in titre if it is to compete with the conventional batch process.

Under anaerobic conditions *Propionibacterium freudenreichii* grows well on mixtures of amino acids (from cornsteep liquor) and monosaccharides to produce cobinamide (factor B), but formation of this intermediate is depressed in the presence of oxygen.

Nevertheless, the same organism is able to convert cobinamide to vitamin B_{12} under aerobic conditions by addition of a nucleotide containing the base 5,6-dimethylbenziminazole (see Chapter 15). Continuous fermentation for production of vitamin B_{12} must therefore be divided into a primary anaerobic stage and a secondary aerobic phase. A suitable nutrient medium is composed of cornsteep liquor equivalent to 0·32%N, glucose 10%, and cobalt (as the chloride) 2·75 μg/ml. The glucose and cobalt chloride are sterilized separately. The optimum pH is 7·0 (regulated by ammonia additions), and the optimum temperature is 30°C. Deoxygenated mash is fed into the first fermenter, and the flow rate is adjusted for a 60 hr retention period. The primary stage is kept strictly anaerobic by bubbling nitrogen through the medium. Overflow from the first stage is pumped to the aerated secondary stage fermenter of similar volume, to give a retention time of 60 hr. To the secondary stage fermenter 5,6-dimethylbenziminazole is added in an amount sufficient to maintain a concentration of 0·01% w/v.

This continuous fermentation process yields approximately 14·5 μg/ml vitamin B_{12} and has been run continuously for nearly 200 days, but it cannot yet compete with the higher yielding (about 20 μg/ml) commercial batch process.

2.5. *Filamentous micro-organisms*

The theory of continuous culture of micro-organisms shows that in the steady stage the dilution rate D is equal to the specific growth rate μ (eq. (19.3)), and the latter has already been shown to be equal to $\log_e 2/t_d$, where t_d is the doubling time (see eq. (19.1) *et seq.*). The dilution rate is the reciprocal of the "mean residence time" (t_R) so that

$$t_R = \frac{t_d}{0{\cdot}693} = 1{\cdot}443\, t_d.$$

Now the maximum specific growth rate μ_m can be found from batch experiments and thus the minimum retention time for the continuous process may be calculated.

This theory applies only to organisms that divide exponentially, and many microbiologists will not admit that filamentous organisms can grow in this manner. It is, however, well established that penicillin (produced by *Penicillium chrysogenum*), streptomycin (produced by *Streptomyces griseus*) and chloramphenicol (produced by *S. venezuelae*) can be produced by continuous culture processes, albeit with lower yields than from the corresponding batch processes: the cultures must therefore approximate to exponential growth and must be in a steady state. Microscopic observation of vigorously growing mycelium of *Penicillium chrysogenum* in submerged culture at pH 6·0–6·5 shows that growth occurs by apical extension and also by repeated branching. When growth is vigorous, multiplication by branching is rapid relative to growth by linear cell division, so that the overall growth of the culture approaches an exponential rate of increase. As the rate of growth falls, branching of the mycelium becomes less frequent so that growth is no longer approximately exponential. Finally, the cell contents begin to vacuolate, leading first to ladder vacuolation and finally to loss of cell contents when the mycelium becomes moribund.

A different type of mycelium is found in the presence of the relatively high concentrations of vegetable oil that are usually present when penicillin is being produced at pH 7–8. At this time there is little branching and the hyphae are relatively short. The apices of the hyphae swell to form spherical bodies, frequently containing oil globules, and these in turn can germinate when adequate nitrogen is supplied. The morphology of the culture then reaches a partial equilibrium, so that the theoretical requirement for a steady state is approached, in which production of spherical bodies and their germination into short, relatively broad hyphae is balanced by lysis of the older mycelium. This is the condition found during the phase of penicillin production. It is clear, therefore, that a continuous process for penicillin production will need at least two stages, the primary growth stage at pH 6·0–6·5 and a secondary penicillin producing stage maintained at pH 7·0–8·0.

Actinomycetes too grow vigorously by branching, and therefore similar considerations permit the theory of continuous culture to be applied to these micro-organisms.

2.6. *Continuous culture of algae*

Batch culture of algal protein has been described in Chapter 8, but the economics of production favour a continuous process.

A large-scale continuous production unit for algal protein has been operated in California. *Chlorella* spp. are grown in long, flat, horizontal, glass-covered tanks, and agitation is provided by recycling the culture. Optimum conditions are maintained by introduction of fresh medium containing carbon dioxide, bound in the form of bicarbonate and carbonate, and nitrogen as nitrate or ammonium salt: the atmosphere above the liquid is maintained at 5% carbon dioxide concentration, and the temperature is kept at 20–25°C. The flow of medium is governed by the density of growth, and this in turn is measured automatically by a photo-electric cell. The limiting substrate is theoretically the amount of light adsorbed by the organism, but in practice the growth rate is a function both of population density and of the incident light intensity, because the penetration of light into the culture depends upon the density of cell growth. A population density of 0·36 g dry weight per litre seems to be about optimal for *C. pyrenoidosa*.

Suggested literature for further reading

1. BARTLETT, M. C. and GERHARDT, P., Continuous antibiotic fermentation: design of a 20 litre single stage pilot plant and trials with two contrasting processes, *J. biochem. microbiol. Technol. Engng.* **1**, 359–77 (1959).
2. BRUNEL, J., PRESCOTT, G. W. and TIFFANY, C. H. (eds.), *The Culturing of Algae: a Symposium*, Charles F. Kettering Foundation, Ohio, U.S.A. 1950.
3. GEIGER, K. H. and COMPTON, J., *Continuous Fermentation Process*, Canadian Patent 545867 (1957).
4. HERBERT, D., ELSWORTH, R. and TELLING, R. C., The continuous culture of bacteria: a theoretical and experimental study, *J. gen. Microbiol.* **14**, 601–22 (1956).

5. MEANS, C. W., SAVAGE, C. M., REUSSER, F. and KOEPSELL, H. J., Design and operation of a pilot plant fermenter for the continuous propagation of filamentous micro-organisms, *Biotechnology and Bioengineering* **4,** 5–17 (1962).
6. MONOD, J., La technique de culture continue theorie et application, *Annls. Inst. Pasteur, Paris* **79,** 390–410 (1950).
7. PIRT, S. J. and CALLOW, D. S., Continuous-flow culture of the filamentous mould *Penicillium chrysogenum* and the control of its morphology, *Nature, Lond.* **184,** 307–10 (1959).
8. SIKYTA, B. and SLEZAK, J., Reproduction of batch growth curves in continuous culture, *Arch. Mikrobiol.* **49** (4) 341–7 (1964).
9. VAN RIJN, L. A., *Improvements in or Relating to the Continuous Fermentation of Liquids*, British Patent 18045 (1906).
10. Continuous culture of micro-organisms, *Soc. Chem. Ind. Monograph*, No. 12, Society of Chemical Industry, 14 Belgrave Square, London (1961).

CHAPTER 20

Industrial Methods for Production of Vaccines

JENNER observed in the late eighteenth century that dairymaids infected with cow-pox seldom suffered the ravages of smallpox in their later years. He subsequently showed in 1796 that children could be protected without serious side-effects against smallpox for a period of several years by inoculation with lymph taken from cow-pox lesions of infected dairymaids or from similar lesions in calves. Refinements of this process are used to this day in the common procedure for vaccination with attenuated strains of live Vaccinia (cow-pox) virus against smallpox.

Protection by related methods against many other human and animal diseases is now possible and the term vaccine has come to mean any product giving immunity to disease.

Immunity is the term used to denote the interaction between host and parasite: its manifestation may range from complete susceptibility to complete resistance of the host to disease.

Acquired specific immunity ordinarily depends on production in the blood of specific globulin molecules (antibodies) as a response to the stimulus of foreign protein or other substance having large molecular weight (antigen) that may have been introduced either by natural infective disease or by artificial administration. Because the tissues of the host play an active part in producing this resistance to disease, the phenomenon is termed active immunity. It is sometimes possible to transmit this resistance to another normally susceptible animal by the injection of immune serum: by this means passive immunity is conferred.

Three types of antigens can be distinguished and industrial methods of production appropriate to each are available.

(1) *Killed vaccines.* These are suspensions of killed bacteria or viruses. Typical products include pertussis or whooping-cough vaccine (*Bordetella pertussis*), typhoid-paratyphoid or T.A.B. vaccine (mixture of selected smooth cultures of *Salmonella typhi*, with *S. paratyphi* A and B), and the Salk poliomyelitis vaccine.

(2) *Bacterial toxoids or inactivated bacterial toxins.* These include tetanus toxoid or ATT (from *Clostridium tetani*) and diphtheria toxoid (from *Corynebacterium diphtheriae*).

(3) *Live attenuated vaccines.* These are suspensions of living attenuated cultures. Products include Vaccinia virus vaccine for vaccination against smallpox, Sabin poliomyelitis vaccine, B.C.G. (Bacille–Calmette–Guérin) vaccine for protection against tuberculosis, and *Brucella abortus* vaccine for immunization of cows against contagious abortion.

1. Vaccine production in aerobic tissue cultures

Viruses require for multiplication the presence of living host cells and suitable tissue can frequently be provided by a "cell line" or a cell population isolated from a single ancestral tissue and maintained *in vitro* by serial sub-cultures. The techniques of tissue culture employed differ little from standard microbiological techniques for culture of nutritionally exacting micro-organisms.

1.1. *Metabolism of cells in culture*

The metabolism of all cell lines is essentially similar, and it is therefore not surprising that they have also common nutritional requirements. Carbohydrates, amino acids, accessory food factors and a physiological balanced mixture of ions, with maintenance of suitable pH and osmotic pressure, must be provided. The necessary salts for maintenance within narrow limits of pH and ionic concentration are chlorides of sodium, potassium, calcium and magnesium, with available phosphate and with bicarbonate to supplement the CO_2 derived from respiration.

Glucose is usually provided as a source of carbon and of energy, but it may sometimes be replaced by fructose, galactose, mannose or ribose. Cells in culture are able to metabolize carbohydrate by all the main pathways that have been described in preceding chapters: both glycolysis and the oxidative pentose-phosphate pathway (see Chapter 9) may be utilized, and the key metabolite, pyruvate, may be converted to lactic acid (see Chapter 10), or may be completely oxidized to CO_2 and water via the citric acid cycle (see Chapter 12) and the respiratory chain of enzymes.

Cells in tissue culture lose some of their ability to synthesize amino acids so that thirteen or sometimes fourteen amino acids must be provided in the medium: valine, leucine, isoleucine, phenylalanine; tryptophan, threonine, cysteine, cystine, methionine and sometimes tyrosine; glutamic acid or glutamine; lysine, arginine, histidine. The requisite accessory food factors to be supplied in culture medium are choline, folic acid, inositol, nicotinamide, pantothenic acid, pyridoxine, riboflavin and thiamin, along with trace quantities of biotin, vitamin B_{12} and the fat-soluble vitamins.

Chemically defined media devoid of protein are necessary when the final vaccine is intended for human use, but supplementation with serum and other biological ingredients is permitted in the early stages. Typical media for growth are mixtures of three components.

(i) Glucose-salt solution (e.g. Hank's solution, Earle's solution, Gey's solution, PBS or phosphate basic solution of Dulbecco: compositions of these media are given in reference 4).

(ii) Serum, preferably from the same species as that from which the cell culture was derived.

(iii) Hydrolysate of animal proteins containing accessory food factors.

In recent years wholly synthetic media have been widely used, especially in the preparation of vaccines for use in medicine: synthetic medium 199 is most favoured. It is prepared in a standardized manner from sixty-two ingredients that are mixed in a given sequence of twenty-four stages (see Reference 9).

Cell lines no longer fulfil their normal physiological roles, and they may also differ both in metabolism and in morphology from the cells in the tissue of origin. Given suitable nutrition and other environmental conditions, cell lines may, in theory, be maintained indefinitely: the cells usually exhibit a tendency towards increased atypical nuclear divisions, so that the chromosome number is no longer constant, and distinction between normal and neoplastic cells may become difficult. It is therefore usual to express the chromosome complement of a cell line in terms of a "modal number" dependent both on the tissue of origin and on the environmental conditions.

1.2. *Production of Salk poliomyelitis vaccine*

Methods used for production of vaccines are basically similar, so that description of the methods used for production of Salk poliomyelitis vaccine should serve as a general example.

Kidneys excised by sterile dissection from rhesus monkeys are washed in sterile Hank's solution, cut into small fragments, washed again thoroughly in saline phosphate buffer and immersed in 0·25% trypsin in phosphate basic solution of Dulbecco (PBS), made alkaline (pH 7·4–8·0) with 2 ml of 1 N.NaOH per litre, before incubation under conditions optimal for action of the enzyme. Trypsinization depolymerizes the intercellular matrix of the tissue fragments and separates the individual but still viable cells forming a suspension that can be preserved by serial transfer at 30°C or by storage at 4°C or in the presence of glycerol at —70°C.

A simple form of apparatus for trypsinization may be constructed from a 1 l. measuring cylinder having two side-arms inserted just above the base and connected by 4 ft of silicone rubber tubing. Synthetic medium 199 at pH 7·2–7·4 containing small pieces (1 mm^3) of cortical kidney tissue and 0·25% trypsin is pumped from the cylinder through one arm, to be recirculated to the cylinder through the other arm: a flow inducer (Watson-Marlow Air Pump Company, Marlow, Buckinghamshire, England) is suitable for this purpose. The larger kidney fragments fall to the bottom of the cylinder and these alone are circulated and disintegrated by the combined action of trypsin and the pump. This technique releases diminutive groups of cells that ascend through the fluid column in the cylinder to float at the surface, where they escape further mechanical disintegration.

The suspension of trypsinized cells is diluted 100- to 500-fold with culture medium before distribution as a shallow layer of fluid in suitable sterile glass containers. The trypsinized cells have an inherent tendency to adhere to a glass surface, but uniform distribution is essential: deposition in stoppered, flat, static, Roux bottles or rotation in stoppered roller-tubes is satisfactory.

Incubation of the cultures for 4–6 days at 37°C permits growth of a confluent and uniform sheet of cells adhering to the glass wall. The culture fluid is removed and must be replaced by synthetic medium 199 or by maintenance medium devoid of protein if the product is intended for use in medicine. Inoculation with live virus and incubation for 4 days at 37°C permits maximum growth of virus accompanied by cytopathic phenomena depending on the tissue used and on the nature of the infesting agent. The suspension of virus particles is harvested, centrifuged to remove cellular debris and filtered to remove large clumps of virus.

The suspension of live virus must then be inactivated, and this is usually done by adding formalin and storage at the required temperature and for the time needed to inactivate the virus without changing its antigenic properties. The inactivating agent must be finally removed or neutralized by appropriate methods before the killed virus is stored at 4°C.

1.3. *Testing for safety and for potency of vaccines*

A large proportion of the cost of manufacture of Salk poliomyelitis vaccine is that of testing for safety and potency. Safety control of killed vaccine is necessary to ensure removal of all active virus. The test for the presence of live virus is done by inoculation into cultures of sensitive cells, but virus-inactivating, toxic substances must first be removed by dialysis with the same synthetic medium as is used for producing the vaccine. The potency is determined by antibody response in experimental animals (guinea pigs, rats, or 1–3-day-old-chicks) to tenfold serial dilutions of the vaccine. The presence of antibodies is detected by permitting samples of blood from the test animals to clot before dilution of the serum with an equal volume of active virus. Antibodies, if present in the serum, will neutralize the virus and prevent its cytopathic action in a suitable sensitive tissue culture. Similar methods are adapted to testing safety and potency of attenuated vaccines and bacterial toxoids.

2. Production of bacterial toxoids in anaerobic submerged fermentations

Certain anaerobic, gram-positive, spore-forming bacteria of the genus *Clostridium* cause disease in man and in animals: these include the causative organisms of gas gangrene (*C. welchii*), tetanus (*C. tetani*) and botulism (*C. botulinum*). They frequently possess both saccharolytic and proteolytic properties, but *C. welchii* is an example of a predominantly carbohydrate-decomposing species and *C. tetani* is an example of a predominantly protein-decomposing species.

Active immunity against these diseases is conferred by injection of appropriate bacterial toxoids. The toxins required for preparation of these products may be produced by anaerobic culture of the appropriate pathogen in industrial fermenters.

Media used for growing *Clostridium* spp. are usually hydrolysates of proteins prepared by partial enzymatic digestion of horse meat and are sterilized by steam at 15 psig. The fermentations are anaerobic so that satisfactory growth and toxin production are easily obtained in deep fermenters (fermenters containing 500 l. of medium are not uncommon) by incubation for up to 6 days (for *C. tetani*) without aeration but with sufficient agitation to maintain temperature and with a flow of nitrogen to flush oxygen from the system. Semi-synthetic media are being used in increasing quantities, because they ease subsequent purification of the toxin.

Selection of a high-yielding strain is of prime importance in toxin manufacture; fermentation conditions merely provide facilities for optimum growth under the requisite anaerobic conditions. It is usual to transfer inoculum at a stage when it is growing vigorously. Samples are taken at intervals from the fermentation broth to test for toxin concentration by reaction with a specific anti-toxin.

At harvest the bulk of bacterial cells is removed by centrifugation in a solids-separating centrifuge, and the supernatant liquid is filtered through a plate-frame filter fitted with asbestos or other microbial filters to ensure complete removal of bacterial cells. The toxin is converted to toxoid by inactivation with formalin: for this purpose the batch is incubated at 37°C until *in vivo* tests show it to be non-toxic. Residual proteinaceous material that may cause unpleasant reaction or even anaphylactic shock in the patient is removed by fractional precipitation with ammonium sulphate at 4°C. The precipitated toxoid is collected, redissolved in buffered saline and dialysed against distilled water at 4°C to remove excess ammonium sulphate. The final product is sterilized by filtration and diluted to final strength with buffered isotonic saline.

The antigenic response of a toxoid can be increased by combination with an adjuvant such as aluminium hydroxide, but this procedure is usually adopted only for veterinary products.

Suggested literature for further reading

1. COOPER, P. D., WILSON, J. N. and BURT, A. M., The bulk growth of animal cells in continuous suspension culture, *J. gen. Microbiol.* **21**, 702–20 (1959).
2. CRUICKSHANK, R. (ed.), Acquired immunity, *Mackie and McCarthy's Handbook of Bacteriology*, 57–71, Livingstone, Edinburgh and London, 1960.
3. PARKER, R. C., *Methods of Tissue Culture*, 3rd edn., Pitman Medical Publishing Co., London, 1961.

4. PENSO, G. and BALDUCCI, D., *Tissue Cultures in Biological Research*, Elsevier, Amsterdam, London and New York, 1963.
5. PERLMAN, D. and GIUFFRE, N. A., Growth of animal cells in tissue culture, *Prog. ind. Microbiol.* **4,** 61–94 (1963).
6. SALK, J. E., YOUNGNER, J. S. and WARD, E. N., Use of color change of phenol red as the indicator in titrating poliomyelitis virus or its antibody in a tissue-culture system, *Am. J. Hyg.* **60,** 214–30 (1954).
7. UNGAR, J., Production of Vaccines, *Research, Lond.* **13,** 304–9 (1960).
8. VAN RAMSHORST, J. D., Potency and purification of tetanus toxoid obtained by means of the Mueller medium, *Antonie van Leeuwenhock* **23,** 97–108 (1957).
9. WILLIAMS, E. N., *Tissue Culture* (Methuen's Monographs on Biological subjects), Methuen, London, 1954.
10. WILLMER, E. N. (ed.), *Cells and Tissues in Culture: Methods, Biology and Physiology* (3 vols.), Academic Press, New York and London, 1964–5.
11. Attenuated poliomyelitis vaccine, *Lancet* II (7366) 946–8 (1964).

CHAPTER 21

Present Trends and Possible Future Developments in Industrial Microbiology

MICROBIOLOGY can never be an exact science, so that the scientist with an artistic feeling for microbial processes is at a premium, but knowledge of the underlying biochemistry and chemical engineering is evolving so rapidly that the time is not far distant when an industrialist clinging to the artistry that has served him so well in the past will no longer be able to ensure economic survival.

To consider the future we must first remember the past. Fermentation products that were at their peak between the two World Wars (e.g. acetone, butanol, industrial ethanol) can no longer compete in industrialized countries with the cheaper products from the petrochemical industries. This history suggests that fermentation processes for the production of simple chemical substances are probably doomed to become uneconomic by comparison with the corresponding synthetic processes.

Industrial microbiological processes are much more favourably placed in economic terms when the product is chemically complex or asymmetric in structure (e.g. vitamin B_{12}, griseofulvin, L-2,3-butyleneglycol). It seems probable that industrial microbiology will hold its own for these purposes, and that the development of commercial microbiological processes will continue to keep pace with the establishment of new antibiotics in medicine: microbiological production of cephalosporin C for use in the chemical synthesis of cephaloridin is the most recent example.

The procedures employed in the industrial production of these chemically complex products may be expected to change as the processes become better understood. It is gradually being learned that metabolic products are formed in the "steady state" and that, although continuous fermentation may be economic only in special circumstances, it is possible to direct the course of fermentations in a manner that could not be contemplated in the early days of batch fermentations. The chemical, biological and biochemical monitoring of these controlled fermentations will become more and more amenable to automation by chemical engineers.

Perhaps the brightest future for industrial microbiology may lie in the exploitation of the enzymes that catalyse microbiological reactions. It may be that microbial enzymes, crude forms of which already find employment in food industries, will eventually be stabilized, singly or in multi-enzyme complexes, on adsorption columns, thereby permitting their use as immensely versatile catalysts of difficult or complex industrial chemical reactions. Apart from the development of completely new synthetic processes, these developments may facilitate standardization and simplification of the many important single-step industrial microbiological processes now employed, for example, in the transformation of steroids.

Meanwhile, other industrial processes are likely to remain essentially microbiological. Reduction of B.O.D. in industrial wastes and in sewage and the production of beers and wines by fermentation may be expected to continue, but here, also, technological progress may result in universal adoption of continuous culture processes. It may be that some of the more novel forms for microbiological disposal of waste (e.g. conversion of petroleum industry wastes to protein) may develop into major sources of protein for feeding both man and his domesticated animals.

Finally, we might expect vaccine production to become generally feasible in laboratory-scale, continuous single cell-line tissue cultures, and this may lead, under some special conditions, to techniques for vaccine production in submerged aerobic tissue cultures.

Index